ATM TECHNOLOGY: AN INTRODUCTION

Marc Boisseau

Michel Demange

Jean-Marie Munier

INTERNATIONAL THOMSON COMPUTER PRESS

I(T)P An International Thomson Publishing Company

London • Bonn • Johannesburg • Madrid • Melbourne • Mexico City • New York • Paris
Singapore • Tokyo • Toronto • Albany, NY • Belmont, CA • Cincinnati, OH • Detroit, MI

ATM Technology: An Introduction
Copyright © 1996 International Thomson Publishing
Copyright © 1994 Éditions Eyrolles, Paris

I(T)P   A division of International Thomson Publishing Inc.
The ITP logo is a trademark under licence

For more information, contact:

International Thomson Computer Press
Berkshire House
168–173 High Holborn
London WC1V 7AA
UK

International Thomson Computer Press
20 Park Plaza
Suite 1001
Boston, MA 02116
USA

Imprints of International Thomson Publishing

International Thomson Publishing GmbH
Königswinterer Straße 418
53227 Bonn
Germany

International Thomson Publishing Asia
221 Henderson Road #05–10
Henderson Building
Singapore 0315

Thomas Nelson Australia
102 Dodds Street
South Melbourne, 3205
Victoria, Austraia

International Thomson Publishing Japan
Hirakawacho Kyowa Building, 3F
2-2-1 Hirakawacho
Chiyoda-ku, 102 Tokyo, Japan

Nelson Canada
1120 Birchmount Road
Scarborough, Ontario
Canada M1K 5G4

International Thomson Editores
Campos Eliseos 385, Piso 7
Col. Polenco
11560 Mexico D. F. Mexico

International Thomson Publishing South Africa
PO Box 2459
Halfway House
1685 South Africa

International Thomson Publishing France
1, rue St. Georges
75009 Paris
France

British Library Cataloguing-in-Publication Data
A catalogue record for this book is available from the British Library

Library of Congress Cataloguing-in-Publication Data
A catalog record for this book is available from the Library of Congress

ISBN 1-85032-304-6

Commissioning Editor Liz Israel Oppedijk
Typeset by On Screen, West Hanney, Oxfordshire, UK
Printed in the UK by Clays Ltd, St Ives plc

Contents

Foreword

The 1990s is the decade in which telecommunications will take their revenge on computing. In today's world, telecommunications have become as indispensable as computing, yet their technical progress has been slower: whereas the power of computers has doubled every two years, it has taken the public service network more than 20 years to advance from several kilobits a second to a hundred kilobits a second. However, in the next few years, running up to the millennium, the situation will be reversed: while progress in computing has been slowed by the recession, advances in telecommunications have undergone an extraordinary acceleration, whose effect will be to increase network performance by a factor of several hundred, if not thousand, in just a few years. This spectacular progress is based on advances in key technologies in two areas: physical telecommunication devices – fibre optics, Hertzian or infra-red channels, ultra-high frequency components, either hybrid or opto-electronic – and logical structures – the architectures and protocols required for managing these speeds and integrating information flows with very different characteristics, such as voice, data and video.

Marc Boisseau, Michel Demange and Jean-Marie Munier's book invites us to explore this second area by describing the results of ten years of international research aimed at defining a single solution to the problems of transmitting, switching and multiplexing heterogeneous flows on broadband networks in an economically viable and internationally acceptable way. It is worth underlining the fact that Europe has played a dominant role in the development of ATM (Asynchronous Transfer Mode), starting with the original research at CNET on asynchronous time multiplexing and continuing with the projects in the European Community's RACE programme right up to the present-day experiments on the national and international ATM networks in Germany, Spain, France, Italy, Great Britain and Sweden.

This book explains the various parts of ATM – architecture, flow control, error handling, performance measurement – and describes how the ATM techniques are applied when creating broadband integrated networks, local area networks and exchanges. It is aimed at both the reader who wants a rapid understanding of the basic principles and someone who wants to know the details of the cell formats or the synchronization algorithm. The authors' training, in one of the great telecommunications research laboratories, uniquely equips them for explaining the reasons behind the technical decisions taken by ATM's designers, sometimes not without a certain sense of humour, such as when they reveal that the ATM cell's payload is 48 bytes because 48 is the arithmetic average between 64, proposed by the Americans, and 32, supported by the Europeans. And they do not hesitate to point out the problems that are yet to be solved and the options between which choices will have to be made.

For the first time in many years, a standard in which Europe has played a major part is beginning to impose itself on the worldwide information technology and telecommunications industries. The important thing now is that European industries and economies should be the first to benefit from the intellectual investment made by our researchers, and, in order for this to happen, that ATM should spread rapidly throughout Europe: in higher education, and amongst manufacturers, operators and users. No doubt this comprehensive, clear and precise book will play its part.

Jean-Jacques Duby
Scientific Director
Union des Assurances de Paris

Preface

ATM (asynchronous transfer mode) is a switching, multiplexing and transmission technique that is a variation on packet switching in so far as it uses short, fixed-length packets called cells. The handling of the cells in the switching units is limited to analysing their headers so that they can be routed to the appropriate queues. The flow control and error-handling functions are not carried out in the ATM network, but are left to the user applications or the access devices.

Because of these characteristics, ATM can respond reasonably to the constraints imposed by traffic as different as voice, moving images or data. This universal transfer mode makes it possible to integrate all types of services on a single network access. Although it was first designed as the technical solution for broadband public networks, ATM is also becoming the technology for future private networks and local area networks.

This book is divided into five chapters and an appendix:

- Chapter 1, Switching techniques (p. 1), explains the reasons behind the choice of ATM and the resulting possibilities. It is not essential to read this chapter to understand the rest of the book.
- Chapter 2, Cell relay (p. 17), describes the functions of the ATM layer (routing, multiplexing cells), the underlying physical layers and the layers for adapting cells to various types of application (AALs ATM adaptation layers).
- Chapter 3, ATM switching units (p. 59), recalls the principles of conventional switching modes and describes the functions of an ATM switching unit. It describes the various techniques for storing cells as well as the types of switching devices.

- Chapter 4, Broadband ISDN (p. 71), describes the target application for ATM technology in wide area public networks. It also describes the deployment of ATM technology in Europe and the USA.
- Chapter 5, ATM and private networks (p. 89), shows how ATM technology can also be integrated into wide-area private networks and local area networks.
- Appendix A, ATM standardization (p. 113), describes the work undertaken by the main standardization bodies.
- Appendix B (p. 121) briefly describes the families of ATM products and gives their main characteristics.

1 Switching techniques

1.1 A historical perspective

The technological need for high-speed networks is a result of the considerable progress made in computing in the past 10 years. This progress has been exemplified by two major changes:

- the change from text-based to image-based displays;
- the distribution of processing power and data storage.

For telecommunications networks, these two changes imply high bit rates (an image contains at least 10 times more basic data items than a text) and extremely short routing delays if there is to be no impediment to the distribution of computing power and data storage.
One simple way of estimating the needs, in terms of communication capacity, resulting from these changes, is to correlate two empirical laws:

- Joy's law, which states that computing power, expressed in millions of instructions per second (MIPS), doubles every 2 years; and
- Ruge's law, which estimates that the communication capacity necessary for each MIPS is 0.3–1 Mbit/s.

If we reckon that in 1990 each computer had an average computing power of 100 MIPS, communications needs will increase geometrically in the next few years and reach a value of between 300 Mbit/s and 1 Gbit/s before 2000. Even if average needs are probably 10 times less than this, they still largely exceed the capacities of present-day networks, both local and wide area.

High bit rate and short delay are two important characteristics, to which we must add a third – the uniformity of the supporting technology for obvious reasons of both economies of scale and integration. These three criteria formed the objectives of studies carried out in the 1980s.

All the studies assumed that flow control and error-handling functions would be dealt with outside the network. This hypothesis was based on the high quality of digital transmission trunks and the inadequacy, at high speed, of protocols that operate hop by hop.

Furthermore, a consensus rapidly emerged around the idea that it should be possible to use a single switching method whatever the type of flow involved.

Frame relay was the first protocol that implemented these principles. It was also at this time that the idea of bridges that forwarded MAC (medium access control) packets was proposed for interconnecting local area networks.

Among these studies, we should mention several important examples: Datakit, AT&T's Voice/Data fast packet switching, IBM's PARIS project and CNET's Prélude. All these projects used packet switching as their basic concept.

Before we describe the techniques that led to the definition of the ATM transfer mode, it would be useful if we specified the characteristics of the two conventional switching modes.

- **Circuit switching** has the advantage of total data transparency; furthermore, it completely satisfies the real-time requirements of voice and video flows. Thus, adapting it to high speeds seemed a possibility. However, the disadvantage of this technique is that it only provides circuits with predetermined bit rates (for example, 64 kbit/s channels in the integrated services digital network). Therefore, it would be necessary to plan for a set of fixed bit rates corresponding to the various projected services. Such planning is difficult, even undesirable, because a given service does not necessarily correspond to a specific bit rate, even if it is only because of the improvement, over time, of data compression algorithms. This avenue of research was abandoned because of its lack of flexibility.

- **Packet switching**, based on the notion of the virtual circuit, could provide that adaptability, and make it possible to use communication links efficiently. Frame relay had already shown that communication protocols could be simplified, but whether this principle could be used for flows other than data still needed to be proved. In particular, it still needed to be demonstrated that such a technique could emulate the characteristics of a circuit.

Two avenues of research, described below, came to similar conclusions:

- ATD (asynchronous time division);
- FPS (fast packet switching).

1.1.1 The ATD technique

This technique used very short fixed-length packets (about 16 bytes) with a header limited to 3 bytes containing a label for a virtual circuit type routing. The use of short packets guarantees, as in the case of circuit switching, a brief and relatively constant delay, which means that, for example, voice signals can be transferred without using echo cancellers.

This ATD technique was mainly promoted by European organizations (manufacturers, operators, the RACE project) in the context of studies and models based, for the most part, on isochronous flow transfer (voice and video). Furthermore, these projects made no assumptions about a particular transmission infrastructure. In particular, the aim of research by CNET at Lannion was to provide high-speed circuits for video communication in a residential context, as well as data transmission and high-quality sound. The Prelude experimental network, based on a basic switching matrix with 16 incoming lines and 16 outgoing lines working at 280 Mbit/s, was used to evaluate the soundness of the chosen approach: a suitable packet mode can carry traffic with differing characteristics, including isochronous flows.

In their first phase, the Bigfon and Berkom experimental networks in Germany were based on a circuit-switching technique. However, the growing consensus around ATD caused the Deutsche Bundespost to delay installing its public broadband network, and, in 1988, Siemens became the first manufacturer to install an experimental exchange, for Berkom, based on these principles.

1.1.2 The FPS technique

At the same time, research was being undertaken by organizations such as AT&T Bell Labs, Network Systems, Bellcore, GTE Labs and IBM (PARIS experimental network), with the main aim of transmitting computer data efficiently at very high speeds. These studies were concomitant with the development of the SONET (*synchronous optical network*) standard for networks based on fibre optics, which were beginning to be installed in the USA.

The term 'fast packet switching', popularized by J. Turner, covers most of these projects. They were based on short packets (about 100 bytes), of either fixed length (Bellcore) or variable length (AT&T, IBM). Quite a large header (about 5 bytes) contained, along with the label, binary elements used to distinguish between different levels of priority. The performance they were aiming for meant that the traditional packet-switching protocols had to be simplified and the switching functions carried out by hardware components.

1.1.3 Fast packet switching

The two avenues of research described above have important characteristics in common.

- The network access supports all types of traffic: voice, data, fixed and moving images.
- The transfer mode to the access is flexible and allows for dynamic allocation of the bandwidth according to the immediate needs of the user system.
- Statistical multiplexing of the high-speed digital links is suitable for bursty traffic. For the user, it results in lower costs, whereas for the network provider, it means that the use of links can be optimized. Individual end-to-end users may use a sig-

nificant proportion, or even the whole, of the link's capacity for limited periods.

- Asynchronous transfer is particularly adapted to variable-speed coding. In this context, we should note that, even though they have traditionally been transmitted as continuous flows, voice and video are, by nature, bursty. In the case of voice, an activity detection mechanism can be used to block the coder when the sound source is silent; by compressing the vocal signal during activity periods we get an average flow rate of 10 kbit/s. Similarly, a video source is extremely bursty: when movements in the image are slight, there is little difference between successive images; the new information is limited and can be transmitted infrequently, in the form of packets. Conversely, if there is a rapid movement or a complete change of image, the new information increases considerably, and the source then transmits a burst of packets at a much higher rate. Obviously, this form of coding optimizes the use of the transmission lines, without complicating the operation of the source.

- The transfer mode is unique. Even the most enthusiastic proponents of circuit switching did not manage to propose it as the sole switching technique, and there were several hybrid approaches: circuit switching for continuous flows, packet switching for bursty traffic. The universality of the proposed transfer mode is a most important advantage.

All these common characteristics were recognized by the packet switching lobby, but some questions still had to be resolved: the size of packets, fixed or variable length, and so on. Furthermore, the use of circuit switching had not been excluded. It was within the ITU-T that these crucial questions were resolved.

1.1.4 The role of the ITU-T

In June 1985, CCITT Study Group XVIII formed a Broadband Task Group (BBTG) responsible for looking at matters concerning the user interface of broadband ISDN. The CCITT is now called **ITU-T** (International Telecommunication Union – Telecommunication Standardization Sector).

STM : Synchronous transfer mode

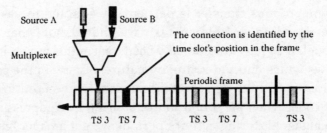

ATM : Asynchronous transfer mode

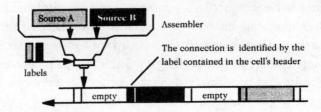

Figure 1.1 Comparison between synchronous and asynchronous transfer modes

Roughly speaking, the two propositions illustrated in Figure 1.1 were in competition for international standardization.

- The first was a synchronous approach, called STM (synchronous transfer mode) by the ITU-T, based on rapid circuit switching. On a multiplexed link, an STM channel is identified in a frame by the position of the time slots assigned to it. Fixed-speed channels were proposed: narrow-band ISDN channels and others that could be used for transmitting video signals and high-quality sound. As the SDH (synchronous digital hierarchy) standard was being adopted at the same time, some of the channels proposed could be carried in the payload of future synchronous transmission systems. However, an interface constructed on a fixed-channel structure lacks flexibility: it freezes the characteristics of the services it carries; these services may vary from one country to another, from one client to another, and, above all, they may well change over time.

- An asynchronous approach was proposed by both the supporters of ATD and those of fast packet switching. The ITU-T named it ATM even though the size of the packets (later called cells) had not been decided. The ATM approach does not require a framed transmission system; a connection is identified by the **label** contained in the cell's header. Thus, several connections can be multiplexed at the same time on a '**labelled multiplex**' link. The use of unframed media is made possible by a cell self-delineation mechanism: the system is then called 'pure ATM', referred to these days as 'cell-based systems' (p. 36). Naturally, ATM flows can be carried by framed transmission systems, in particular in SDH containers (p. 35).

1.1.5 The ATM compromise

The ATM transfer mode, which resulted from the studies mentioned above, combines the advantages of previous techniques. Figure 1.2 summarizes the main characteristics of the traditional switching methods (circuit mode and packet mode).

The ATM transfer mode preserves the following advantages of circuit mode.

- The network transmits the cell's **payload** in a totally **transparent** way, just like the eight bits in a time slot.
- The payload has a **fixed length**, just like a time slot's byte, which means that relatively simple high-performance switching mechanisms can be designed.
- The payload is **short**. This characteristic means that a circuit can be emulated, while at the same time guaranteeing a jitter compatible with the constraints imposed by voice or moving image transmission.

Other advantages are preserved from packet mode.

- The source and the network are not tied by the necessity of transmitting and receiving a quantity of information synchronously

Constraints	Circuit switching (ISDN)	Packet switching (X.25)	Cell switching (ATM)
Real time	Yes	No	Yes
Transparency	Yes	No	Yes
End-to-end protocol	Yes	No	Yes
Variable bit rate	No	Yes	Yes
Statistical multiplexing	No	Yes	Yes

Figure 1.2 Criteria for choosing cell switching

with a frame and during a time slot assigned to the connection involved. The exchange with the network is **asynchronous** and the source alone is responsible for its bit rate, within the limits of the contract defined at the start of the communication (**bandwidth on demand**).

- To optimize use of the network, connections can be **statistically multiplexed**, on condition that the quality of service required for each of them allows it.
- Routing by **labels** opens the door to numerous possibilities, such as broadcasting, setting up user groups or dividing the network into a hierarchy of virtual paths and virtual channels.

As shown in Figure 1.3, ATM combines the simplicity of circuit switching with the flexibility of packet switching.

The 48-byte size of the ATM cell's payload is the result of a compromise. During an ITU-T meeting in Geneva, the representatives of the USA and a few other countries recommended a 64-byte data field, whereas the European countries favoured a 32-byte field. As there was no irrefutable technical consensus on the matter, the compromise decision was to adopt the halfway position between the two propositions.

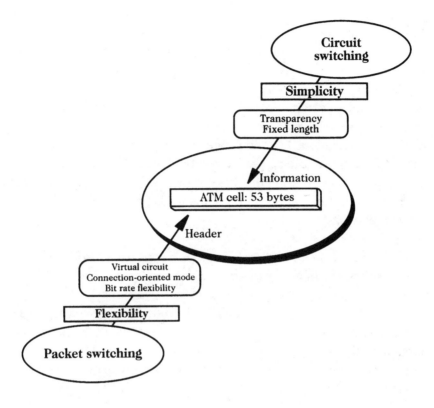

Figure 1.3 The position of cell relay

While on this subject we should note that, when transmitting a digital telephone signal, assembly into cells leads to a delay of 6 ms (48 bytes transmitted at 64 kbit/s), or the equivalent of the propagation time over a distance of more than 1000 km. In medium-sized countries (for example, European countries), this additional delay may require the installation of echo cancellation systems, whereas nowadays these are only required for international connections.

■ 1.2 The development of telecommunications

The ATM concept was only able to take shape and gain the consent of the majority so rapidly because its roots lie deep in general developments in the telecommunications domain. The choices made were not an upset, but rather an integration of the progress made in existing techniques; in the long term, this should lead to the unification of the transfer modes used by all the devices that belong to the communication media (terminals, local area networks, wide area networks, and so on).

The telecommunications world is constantly developing, with each new technique usually building on the preceding ones. Thus, the digital multiplex hierarchy is based on frequency division multiplexing and the frame relay technique is an improvement on packet switching; similarly, narrow-band ISDN can be seen as the final standardization of a telecommunications infrastructure that has constantly been modernizing itself.

The growth in demand for professional communications has led to two types of network:

- the dedicated public networks (X.25, for example);
- private company networks, built using leased lines.

If we followed this logic, we would end up with a network for each service in the public domain and one for each application in the private domain, with the obvious losses of economies of scale. The availability of a multiservice network technology can only be welcomed, in both the public and private domains.

1.2.1 Technological features

The introduction of **fibre optics** into public and private network infrastructures is becoming widespread. They enable interconnected workstations to operate remotely with very short response times, at the same bit rates and with the same quality as on a computer bus.

This has caused a switchover in the roles of the network and the computer. For a long time, communication between processing units was limited by networks that provided only relatively low bit rates. Networks with fibre optics provide a bandwidth equal, if not superior, to that of the processing units' internal buses or input/output channels. The computer and its peripherals are no longer restricted to a single computer room – the era of distributed processing and the client-server model has dawned.

Although the bandwidth made available by fibre optics is large, it is not infinite, and it will always be relatively expensive if large distances have to be taken into consideration. Furthermore, bandwidth needs to

continue to increase, given, among other things, the growing improvement in quality required by terminal equipment (screen size, number of
pixels per image, number of colours) and the introduction of moving
images. For all these reasons, data compression techniques are necessary.

Another important aspect is the generalization of **digital techniques** in
the transfer mode as well as the processing and the storing of data. Allied
to technological advances in very large-scale integrated (VLSI) circuits,
memories and signal processors, these techniques enable important
progress to be made, in terms of both functions and costs.

Thus, these digital techniques have now made it possible to implement
fast packet switching, using hardware components, on condition that it
is relatively simple and allows only a few options. To simplify the associated protocols, the more complex functions (segmentation, error handling, flow control and so on) are off-loaded onto the terminal devices,
whose capacities for low-cost processing are considerable.

Display techniques derived from televisual processes are increasingly
used on workstations: whereas in the past they were limited to managing
text, spreadsheets or still images, they are slowly allowing multimedia
communication (face-to-face conversation via video windows, sound,
text, data transfer and so on). We should note that progress in data compression means that nowadays the bit rate necessary for a high-definition
television-type video image is only 30 Mbit/s, and for VHS-quality moving images only 1.5 Mbit/s.

1.2.2 Planned applications

Tomorrow's network must support the applications we know today,
those that we are planning and, even more importantly, those that we
have not even imagined. As Figure 1.4 shows, its switching mode must
accommodate a large variety of bit rates (from several kbit/s to several
Mbit/s), constant or bursty flows and different qualities of service (more
or less sensitive to variations in delay or to error rates).

In the business world, the main aim is to provide the same communication possibilities to all users, wherever they are located geographically.
Applications that today are restricted to local area networks may thus
become available remotely: high-speed data transfer, graphic applications, computer-aided design and so on. Rapid access to remote data-
bases or servers allows processing to be distributed. In a way, it is the

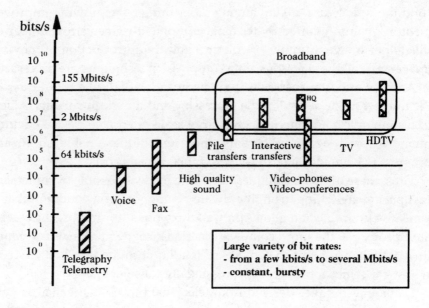

Figure 1.4 The variety of bit rates that need to be supported for the different applications

network, with its distributed servers, that is becoming the company's computer, and thus responsible for response time and quality of service constraints.

In the world of professional applications, there are specific needs awaiting the installation of appropriate networks to become truly available over a wide area: medical imaging with the possibility of immediate annotation, teleconferencing with moving images, remote editing and training, and many more.

It is noticeable that most pilot projects (for example the European RACE projects) involve professional applications.

In the domestic area, the target applications are mainly concerned with entertainment: high-quality television pictures, among others, but also television on request. For this last application, the means for consulting a list of films and performing the classic operations of a local video recorder remotely must be made available. In the years from 1980 to 1985, it was the common belief that the main commercial force behind the success of a broadband network in the residential market would be entertainment. Various difficulties, such as installation of fibre optics that reach homes in the distribution network and competition from other technologies, such as cable and satellite, have pushed that perspective back into the 2000s.

This explains why, faced with a constantly changing competitive market and the need to find new sources of revenue, network operators target the professional market first, and principally the interconnection of local area networks. Paradoxically, the installation of a high-speed network based on fast packet switching relocates the added value to the network's periphery, in the terminals connected to it, which does not necessarily lead to it being adopted rapidly.

The existence of high-capacity digital trunks is essential for all the applications envisaged, whether professional or domestic, both to support the **large flows** generated by the information sources and to guarantee a satisfactory **response time**.

1.2.3 Existing services

Existing services, based on circuit or packet techniques, generally use dedicated networks that are well suited to traditional applications. However, they are only partially satisfactory for the new applications mentioned above, because the latter are multiservice and generate large and bursty flows. Businesses that have to install such applications today have to create private networks, whose backbone is made up of 2.048 or 34.368 Mbit/s lines leased from the operators.

In the long term, the public operators want to provide high-speed services. The technological advances mentioned above have made this possible: end-to-end digitization of the network leading to a very low line error rate, owing to fibre optics being used intensively in the main national and international trunks. Furthermore, standardization makes the mass production of very large-scale integrated circuits feasible, which means that very efficient technologies are available, at a reasonable cost, for switching and signal processing.

In order to take advantage of these technological advances, the architecture of the communication protocols needs to be revised. In particular, the significant reduction in the error rates of transmission systems means that the protocols in the network nodes can be reduced, and the network nodes can be simplified so that they can handle higher bit rates (Figure 1.5).

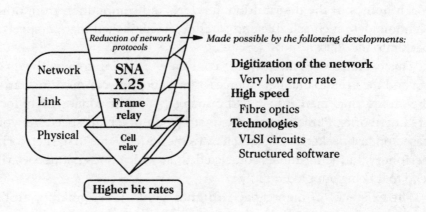

Figure 1.5 Changes in information transfer technologies

1.2.4 Constraints

The objective is to define a single transfer mode, but that mode must be able to satisfy all the constraints imposed by the applications described above. It is not possible to establish a hierarchy of these constraints; they are all equally critical and their relative importance is only a factor of the particular application.

Thus, multimedia applications present numerous constraints that are not necessarily independent (for example the image and the sound must be transmitted with similar characteristics). Furthermore, those constraints may change during a single connection.

Real time

Voice transmission, along with video, is the most common example of an application with real-time requirements: every 125 μs, a sound signal sample is transmitted in the form of a byte, and this must be reconstituted on reception with the same regularity, no matter what happens during transmission. Traditionally, circuit switching provides an isochronous service that satisfies these requirements. However, circuit emulation can be achieved by a packet-type technique, on condition that the global delay and its variations are limited: the use of short fixed-length packets favours such an approach.

Conventional local area networks – CSMA/CD bus, token ring, FDDI (fibre distributed data interface) network – do not guarantee the required real-time characteristics, especially when the network is heavily loaded. FDDI-II extends the FDDI protocol functions in order to respond to these requirements by reserving channels (WBCs, wideband channels) for use by isochronous traffic. Improvements to other local area networks are also being studied (for example, the 100 Mbit/s Ethernet network).

Bit rate

A multiservice network must support a large variety of bit rates; furthermore, these may be bursty or non-symmetrical. On the other hand, the characteristics of a given service and the corresponding bit rate cannot be set once and for all; the speed may vary over time, given the advances made in compression algorithms and the availability of very fast signal processors. Finally, predicting the characteristics of future services is a risky business. Therefore, a multiservice network must be flexible enough to adapt to all sorts of bit rates, if it is to last.

Quality of service

The required quality of service varies according to the application. Some are more tolerant than others of disturbances in the flow of information. A common transfer mode must therefore be reasonably accommodating towards these differences. 'Reasonably' means that a single service cannot be perfect; an adaptation function that can compensate for the network's imperfections must be inserted near the application.

The first supporters of ATM, defined by the ITU-T as the target technology for the future broadband ISDN, were the public network operators and their traditional suppliers. However, around 1991, ATM attracted the attention of the manufacturers of the interconnection devices (concentrators, bridges, routers) and packet switches used in private networks, along with, more recently, the manufacturers of workstations and personal computers, and finally, of software.

Therefore, the first ATM products were put on the market in the private domain, as a supplement to traditional local area network techniques, or even in direct competition with them.

In the longer term, it is possible that this transfer mode will be used end to end between two workstations linked by local area networks or private ATM switching units, which are themselves connected by a public broadband ATM network.

In the next chapter, we present the technical characteristics of ATM not only as they are described in the ITU-T recommendations, but also in the documents approved and published by the ATM Forum (p. 116), a manufacturers' association created in 1991 that now includes some 700 representatives.

2 | Cell relay

2.1 Principles

ATM is both a switching and multiplexing technique and also a transmission technique. It is a variation on packet switching in so far as it only uses short fixed-length packets called cells. Cell handling by an ATM switching unit is limited to analysing the label (part of the header, similar to a logical channel number) so that the cell can be routed to the appropriate output queue. The more complex functions, such as error handling and flow control, are not carried out by the ATM network, but left to the user systems.

These particular features furnish a reasonable solution to the problems posed by the simultaneous constraints of traffic as diverse as voice, moving images and all types of data. Because it is so flexible, ATM can eventually integrate all services onto a common access to a single network.

Cell switching is located between the transmission functions and those concerned with adapting the information flow to the cell format. This gives us a three-layer model (Figure 2.1):

- the ATM layer, responsible for multiplexing and switching the cells;
- the physical layer, which adapts them to the transmission environment;

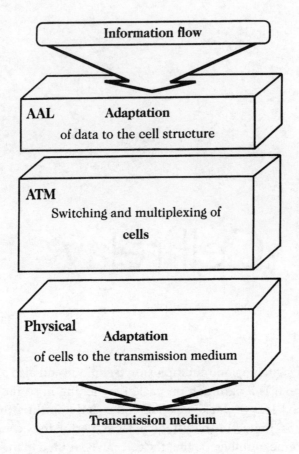

Figure 2.1 Architectural model for cell relay

- the AAL layer (ATM adaptation layer), which adapts the information flows to the cells' structure.

We will discuss these three layers starting with the central one (the cell) and then enlarge the discussion to take in the transmission environment and ending with the adaptation of the information flows.

We should note that cell relay is used not only by ATM technology, but also by DQDB (distributed queue dual bus) technology, adopted by the IEEE as the data transfer and medium access technique for metropolitan area networks.

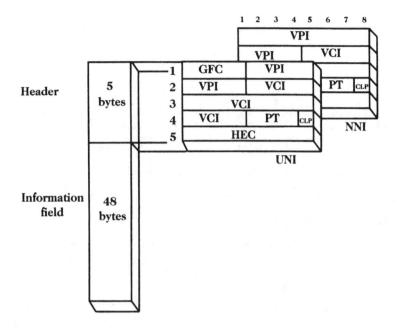

Figure 2.2 ATM cell structure

■ 2.2 ATM layer functions

A cell is 53 bytes long and contains two main fields (Figure 2.2):

- a header (5 bytes), whose main role is to identify the cells that belong to the same connection and allow them to be routed;
- a data field (48 bytes), which contains the payload.

The cells used at the broadband subscriber access (UNI, user network interface) have a slightly different header from those used at the interface between networks (NNI, network node interface).

The ATM cell header used at the interface between the user and the network (UNI cell) contains the following fields:

- a flow control field (GFC, generic flow control), whose definition has not yet been finalized. It is used to regulate priorities and access contentions between several terminals (point-to-

	Type of flow	Congestion indicator	Type of data unit
000	0 User	0 No congestion	0 Type 0 unit
001	0 User	0 No congestion	1 Type 1 unit
010	0 User	1 Congestion encountered	0 Type 0 unit
011	0 User	1 Congestion encountered	1 Type 1 unit
100	1 Network	0 Maintenance (segment by segment)	
101	1 Network	0 Maintenance (end to end)	
110	1 Network	1 Network resources management	
111	1 Network	1 Reserved	

Figure 2.3 Payload type indicator coding (PTI)

multipoint configuration; p. 78). Another proposed use for this field, in the context of local ATM networks, is for flow control and preventing congestion;

- 3 bytes use for the logical identifier (VPI and VCI; p. 21);
- three PTI (payload type identification) bits used to describe the type of payload (user data or network service message; Figure 2.3). In the first case, the last 2 bits provide a congestion indicator, as well as the type of data unit, which is interpreted by the upper layers. This type is transmitted to the AAL by the ATM layer, and can be seen as an extension of the adaptation functions: it is used by the AAL type 5 function to indicate the last cell in a segmentation operation (p. 50);
- one CLP (cell loss priority) bit, whose role is specified on p. 24;
- one HEC (header error control) byte for detecting errors and correcting a simple error involving the header. The handling of this byte, described on p. 30, is the responsibility of the physical layer.

The only difference in the ATM cell header used between networks (NNI cell) is that the GFC field is missing. The corresponding bits are used to extend the logical identifier field.

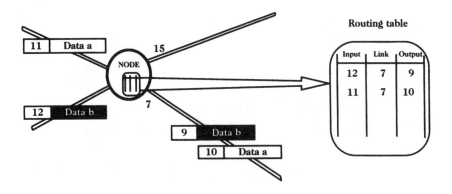

Figure 2.4 Principle of cell routing

2.2.1 Principle of cell routing

Because the ATM service is connection oriented, routing tables are required in the network-switching units. Each cell is set on its route by the intermediate switching units, which associate its identifier with a destination, as shown in Figure 2.4.

As in the case of packet switching or frame relay, the logical identifier only has a local meaning. In this case it is made up of two fields (Figure 2.2):

- a group identifier or VPI (virtual path identifier): it is 8 bits long in a UNI cell and 12 bits long in an NNI cell;
- an identifier of the element in the group or VCI (virtual channel identifier) which is 16 bits long.

A pair composed of a virtual path (VP) and a virtual channel (VC) is the equivalent of a virtual circuit in packet switching or a virtual link in frame relay. The notion of a virtual path is used by the network manager to organize and manage the transmission resources, using permanent or semipermanent virtual links.

As indicated in Figure 2.5, a route is made up of two types of connection: a virtual path connection and a virtual channel connection. Each connection is made up of a concatenation of virtual channels and paths. The hierarchy of identifiers (VPI, VCI) allows for the development of two types of switching units:

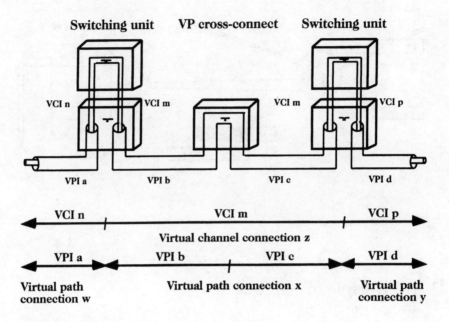

Figure 2.5 Cell relay: the dual routing mechanism

- ATM virtual path-switching units, usually called VP cross-connects (ATM digital cross-connects or ATM DCCs), which use only VPI to forward the data along the route. They are controlled by the network management units;
- ATM virtual channel switching units, which take both identifiers into account (VPI and VCI). These are mainly network access exchanges operated on a call-by-call basis by call-handling mechanisms.

A VP cross-connect is used to route all the virtual channels belonging to the same path as a block. These cross-connects can be used to configure leased line networks, to provide stand-by routes, to make up the interconnection between switching nodes in a connectionless service and so on.

Cells are assigned to a connection according to the source's activity and the network availability. There are two modes for assigning connections:

- permanent assignment or virtual permanent connection, which is the result of a service contract between the network operator and the user;
- assignment on request, call by call, or switched virtual connection, which requires a signalling protocol between the user terminal and the network access switching unit.

This signalling protocol is itself carried on a separate virtual connection which, like any other virtual connection, can be assigned to the signalling activity permanently or on request (call by call). In the latter case, signalling virtual connection set-up makes use of a special procedure, called metasignalling. The protocols necessary for assigning virtual connections on request, call by call, are discussed on p. 81.

2.2.2 Protection against congestion

The mechanisms to be used to avoid network congestion depend on the type of service provided by the ATM layer. The ATM Forum has defined several categories of service, also called QoS (quality of service). The following simplified list gives the main attributes of these categories:

- CBR (constant bit rate) traffic for continuous bit rate applications. A CBR flow induces strong constraints in terms of guaranteed bandwidth and time-dependent parameters such as delay and cell jitter. Circuit emulation is possible with this quality of service. The bandwidth to be reserved is a function of the PCR (peak cell rate), the definition of which is given below (see p. 25).
- VBR (variable bit rate) traffic for bursty bit rate applications. A VBR flow is less demanding than CBR on guaranteed bandwidth and time-dependent parameters. This allows a bandwidth to be reserved depending on the PCR and the SCR (sustainable cell rate), the latter being defined on p. 26.
- ABR (available bit rate) traffic for applications without stringent constraints, able to control their data flow and adapt their instantaneous rate to a value varying between a minimum cell rate (MCR) and the peak cell rate (PCR). Special RM (resource management) cells periodically provide the state of congestion

in the network and allow the stations to increase or decrease their instantaneous rate (see p. 28). For ABR traffic, the bandwidth to be reserved corresponds to the MCR. Although cell delay and jitter obviously cannot be specified, the cell loss ratio has to be minimized.

- UBR (unspecified bit rate) traffic, for which the network does not guarantee any delay, jitter or cell loss ratio. There is no constraint for the stations in sending information, but the network does not reserve any bandwidth for this type of traffic, also called **best effort**.

Table 2.1 summarizes and simplifies the current definitions proposed by the ATM Forum, which in fact splits the VBR category into rt-VBR (real-time VBR) and nrt-VBR (non-real-time VBR). As far as ITU-T is concerned, service classes that are similar to the QoS categories described above have been defined. ITU-T also proposes another type of service, called ABT (ATM block transfer), in which all the cells in the same frame are sent as a block.

In the case of services that require a bandwidth reservation, mainly for CBR or VBR traffic, the terminal devices are responsible for the flows they generate, in accordance with a contract between the user and the network. The **traffic contract** describes the characteristics of the source's traffic, such as its average bit rate, its maximum bit rate and the duration of bursts (**burstiness**). It also defines the quality of service attributes associated with the connection, in particular the CLR, the CTD and the CDV. Depending on the mode used to assign the connection (either permanent or switched), these characteristics can be defined when the subscription is taken out or negotiated call by call.

In the case of UBR service, for which there is no contract, the network does not reserve any bandwidth. End-to-end flow control mechanisms are then needed to adapt the source's rate to the changing capabilities of the network, while minimizing the cell loss ratio.

Despite these precautions, the network can enter a congestion state. The CLP bit in each cell's header is used in the congestion protection mechanisms. This bit can be controlled by the source, which determines the relative importance of the data carried in each cell.

An example of a possible use is in multirate coding of video: the CLP bit is set to 1 in cells that are carrying less important information. If the network becomes congested, these cells are discarded first, leading

Table 2.1 Quality of service

Quality of service	CBR	VBR	ABR	UBR
Specified characteristics				
CDV (cell delay variation)	Yes	Yes	No	No
CTD (cell transfer delay)	Yes	Yes	No	No
CLR (cell loss ratio)	Yes	Yes	Yes	No
Parameters used for bandwidth reservation	PCR	PCR, SCR	MCR	None

to a loss of image quality, which is less important than a loss of transmission.

Protection against congestion calls on a set of mechanisms that are brought into play during different phases of the connection.

Admission control

The first preventive measure is a CAC (connection admission control) for new connections. Given the traffic characteristics, the network must usually decide if it has the necessary resources available to guarantee the quality of service requested by the user, and then reserve those resources for the duration of the connection.

- If the quality of service is very demanding, the bandwidth reserved must correspond to the source's maximum bit rate, also called the peak rate (PCR). Thus, the network can guarantee to route all bursts of information to their destination within a guaranteed time limit and with a practically zero probability of cell loss. The PCR of an ATM connection is expressed in cells per second. It is equal to the inverse of the smallest time interval t (called peak emission interval) separating consecutive cells (Figure 2.6).
- If the quality of service is less stringent, in terms of delay variation and loss of data, the network can take statistical

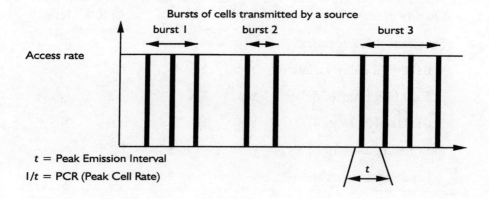

Figure 2.6 Characteristics of the bit rate of a source of traffic

probabilities into account and only reserve a bandwidth that depends of the PCR and the sustainable rate (SCR) for the connection. This latter bit rate is in between the source's average bit rate and its maximum bit rate; the more long bursts the traffic contains, the nearer it will be to the maximum bit rate.

Spacing and policing

Once the connection has been set up, the source implements a spacing mechanism (**source shaping**) when transmitting information to respect its traffic contract.

We should note that, in the case of an access to a network using circuit mode, also called STM, the source could use only the bit rate corresponding to the time slots assigned to it when the connection is established; the network would only take into account the data contained in those time slots, and therefore the bit rate accepted would be **calibrated** by construction. In contrast, when accessing a network using ATM, there is nothing to stop the source offering traffic greater than that specified in its contract on the virtual channel assigned to it, either intentionally or following an equipment failure. Thus, the network must carry out the calibration itself, by controlling and regulating the bit rate offered on the virtual channel in accordance with the contract. It is essential that the network protects itself against such situations; otherwise it will not be able to guarantee the quality of service offered to the connections because there will be overflows of the queues in the switching units.

Therefore, **policing** is implemented at the network access. It is beyond the scope of this book to discuss these control mechanisms in detail (UPC, usage parameter control), as they are complex and still under discussion. For each connection, they usually combine the following functions:

- a **measurement** of the bit rate offered, which usually uses **leaky bucket**-type algorithms whose parameters depend on the quality of service associated with the connection;
- a **control**, which is used to eliminate cells in excess of the contract, or to reduce their level of priority. This last method goes by the name of **violation tagging**.

Spacing can be carried out on the cells before they are fed onto the network, according to the bandwidth reserved for the connection (Figure 2.7). At the expense of an additional delay, such a spacing function ensures an efficient protection of the network.

These policing and spacing functions can be implemented separately on the priority cells (CLP = 0) and on all the cells related to the connection (CLP = 0 and CLP = 1).

Congestion notification

In spite of these precautions taken at the interfaces of the network, a state of congestion can temporarily occur in a switching unit because of the statistical accumulation of bursts of cells belonging to different connections. Such congestion is generally detected by an overflow of a threshold in the switching unit's queues.

An EFCI (explicit forward congestion indication) may then be activated in the cells that have passed through the congested switching unit. This congestion notification is provided by the second bit of the cell header's PTI field (p. 20). Given the projected bit rates for high-speed networks such as broadband ISDN (p. 71), the effectiveness of such a mechanism is not proven. The propagation time between source and receiver is very long compared with the time it takes to transmit a cell (2.73 µs for an access rate of 155.520 Mbit/s, or the equivalent of propagation over 500 m of cable), and a great number of cells can be transmitted before the

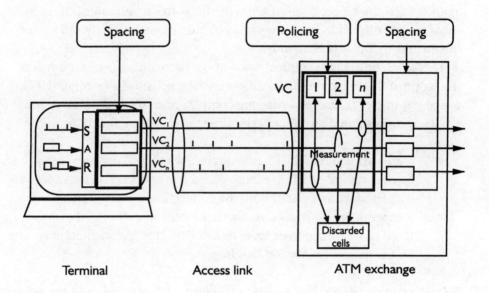

Figure 2.7 Spacing and policing

congestion notification is received. Furthermore, the congestion may then have disappeared.

In the case of ABR traffic, special RM cells are periodically inserted into the source's flow (every 32 user's cells or every 100 ms). Among other information, these indicate the source's instantaneous bit rate, maximum rate allowed and expected minimum rate. When these cells reach their destination, the latter sends them back towards the source and inserts, if needed, congestion information if the header of one or more user's cells received previously contained an active EFCI bit. While receiving this information, the source decreases its instantaneous bit rate. In addition, depending on their state of congestion, the switching units that forward the RM cells may explicitly indicate a maximum allowable value for the bit rate. This rate control mechanism allows the source to use the network resources to the full, with the rate varying between the minimum expected rate and the peak cell rate.

It should be noted that studies are still carried out on other methods for traffic control, such as step-by-step credit-based mechanisms.

Discarding

If the congestion persists or gets worse, the only solution left is for the network to eliminate cells. This operation is carried out in two stages: starting at a certain congestion threshold, only cells whose CLP bit = 1 are destroyed by the congested switching unit; beyond another threshold, all excess cells are eliminated.

Another technique, more efficient but still under study, could limit retransmission of data frames. In the case of UBR- or ABR-type traffic used on local area networks, a congested switching unit could then eliminate complete MAC frames (**frame discard**).

2.2.3 Multiplexing information flows

Cell relay is not only a switching technique, but also a multiplexing technique.

The cells are generated on demand, depending on the source's bit rate. The data items are first adjusted to the size of the payload and the connection's header added. This cell generation process is solely governed by the source's own bit rate, and is not linked to the characteristics of the underlying transmission medium (bit rate, any delineation pattern and so on). This is why the technology is called **asynchronous transfer**, as opposed to circuit switching and multiplexing: in this latter mode (synchronous transfer), the source must provide a data item for each frame, usually every 125 μs to fill the time slot it has been allocated.

The multiplexing of cells transmitted by different sources that share the same access link is similar to the multiplexing of packets belonging to different virtual circuits in packet switching. The discontinuous flow of cells resulting from the multiplexing of several connections is transmitted to the physical layer.

▓ 2.3 Functions of the physical layer

One result of the functional layer structure is that the constraints imposed on the physical layer by the ATM layer are very limited. The flow of cells generated by the ATM layer can be carried in the payload of practically any digital transmission system, which means that it can be adapted to any present or future transmission system.

The physical layer can be divided into two sublayers, which provide the main functions listed:

- the convergence sublayer handles bit rate adaptation, header protection, cell delineation and adaptation to the physical medium's structure;
- the physical medium sublayer is responsible for coding, decoding, scrambling and adaptation to the medium.

2.3.1 Rate adaptation

Usually, the bit rate of the flow of multiplexed cells provided by the ATM layer is not equal to the working bit rate of the physical access link. Rate adaptation, often called **stuffing** or **justifying**, is necessary. The different ways of carrying out this adaptation can be grouped into three main techniques, the third of which is really a combination of the first two.

- To generate a continuous flow of cells, empty cells are inserted into the flow. In the case of a framed transmission system, the resulting flow then corresponds to the transmission link's payload (for example SDH synchronous frames), whereas, if the transmission link is 'cell based', it is equal to the total bit rate of the link. This insertion method has been retained by the ITU-T for broadband ISDN.
- Conversely, the flow of cells can remain discontinuous. This type of flow is mainly found in ATM local area networks, which have yet to be standardized. As the time interval between cells may be of any length, stuffing characters (**idle** symbols) can be inserted to adapt the bit rate. For example, this technique is used for ATM transmission on an infrastructure using the FDDI physical layer at 100 Mbit/s (p. 37).
- A combination of the two previous methods consists in grouping a constant number of cells in blocks, which may be padded with empty cells. The difference between these blocks and the bit rate may be filled by a variable number of stuffing bytes, so as to guarantee a strict sequence of blocks, one every 125 µs. This last method is used for ATM transmission on plesiochronous links, PDH (plesiochronous digital hierarchy).

2.3.2 Header protection by the HEC

As cell routing is based on the VPI and VCI fields, they must be protected because, if there is an error, routing becomes impossible. The HEC provides this protection. Generally, transmission errors are independent from each other, especially in an optical network: they mostly cause isolated errors, which can be corrected relatively simply. In the case of error bursts (due, for example, to configuration modification operations on a network made up of redundant links), correction is not allowed and faulty cells found during that period are discarded. In any case, whether the error is isolated or grouped, if it exceeds the HEC field's correction capacity, the cell will be destroyed.

The receiver has a correction mode and a detection mode for the header protection mechanism.

- In correction mode, which is the normal operating mode, cells whose HEC shows no error syndrome are passed to the higher layers; cells whose HEC shows a single error are passed after the faulty header has been corrected, whereas those with multiple errors are destroyed.
- The detection of an invalid HEC (single or multiple errors) causes a change to detection mode, in which all the cells whose HEC is faulty are destroyed. Conversely, the detection of a cell with a correct HEC causes a return to correction mode.

Figure 2.8 illustrates this two-mode procedure, which is used to protect the receiver as far as possible against bursts of errors: when one error is found, the receiver assumes it is the start of a burst and changes to detection mode, in order to avoid making unnecessary corrections. If this hypothesis is proved false when the next cell is received, it returns to correction mode.

Mathematically, the HEC's value for any given header is derived by applying the following procedure.

- The 32 bits of the header's first 4 bytes are used as the coefficients of a 31-degree polynomial $M(x)$ (the first bit corresponds to term x^{31} and the last to term x^0).
- The polynomial $M(x)$ is multiplied by x^8 then divided (modulo 2) by a generator polynomial $G(x) = x^8 + x^2 x + 1$.

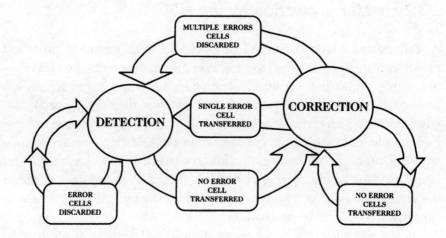

Figure 2.8 Error detection and correction using the HEC field

- The polynomial $C(x) = x^6 + x^4 + x^2 + 1$ is added (modulo 2) to the remainder of the division, thus producing a polynomial $R(x)$ whose coefficients form the 8-bit sequence of the HEC (fifth byte in the header).
- The byte thus generated provides a Hamming distance of 4, a property that allows all single-bit errors to be corrected (only 1 bit in error) as well as the detection of all 2-bit errors.

2.3.3 Cell delineation

When cells are received, their limits have to be identified. This delineation function can be carried out in various ways, depending on the technique used for adapting the bit rate.

Adaptation by continuous flow of cells

In the case of a continuous flow of cells, a delineation function, independent of the transmission system, has been defined. It is based on the detection of the cell header's HEC field. Use of a scrambler improves security and robustness (p. 38).

Cell delineation is based on the use of the HEC field, which protects the first 4 bytes of the header. This method does not rely on any special delineation pattern, as using the HEC allows self-delineation of the ATM cells.

The limits of a cell in a continuous digital flow are detected by determining the byte for which the HEC coding rules are proved to be true (see above). This self-delineation procedure is based on the finite state diagram shown in Figure 2.9.

In the **search** for cell limits state, the mechanism checks, bit by bit, if the HEC coding rules apply to the assumed header. On detecting the first cell limit determined by receiving a valid HEC, the mechanism changes to the **presynchronization** state. In this state, the mechanism continues to check that the HEC coding rules apply to the bytes assumed to contain an HEC.

When *delta* consecutive valid HECs are detected, the change to the **synchronization** state is authorized, whereas the detection of an invalid HEC causes a return to the search state. In the synchronization state, the mechanism continues to check the HECs: when *alpha* consecutive invalid HECs are detected, there is a return to the search state.

The larger the number of consecutive invalid HECs necessary to consider delineation lost, *alpha*, the smaller the probability of losing delineation when there is a transmission error. But the larger the *alpha*, the longer it takes to detect a loss of delineation.

The probability of incorrect delineation diminishes with large values of *delta*, the number of consecutive valid HECs necessary to consider the delineation acquired. However, the larger the *delta*, the longer the delineation acquisition time.

Two sets of values have been proposed according to the type of physical medium:

- 7 for *alpha* and 6 for *delta* for synchronous transmission systems;
- 7 for *alpha* and 8 for *delta* for cell-based transmission systems.

For error rates of 10^{-6} bits, the time between two losses of delineation is larger than 10^{30} cells.

Other rate adaptation techniques

When rate adaptation is not based on the generation of a continuous flow of cells, the delineation technique that uses HEC field detection is not

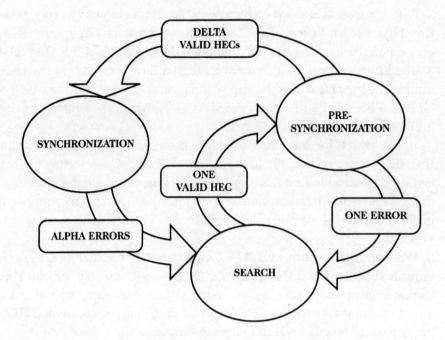

Figure 2.9 Principle of ATM cell delineation

necessarily feasible. In this case, a delineation pattern must be provided for each cell, in the form of bytes or symbols. Some specific examples are described in the next section.

2.3.4 Adaptation to the transmission system

Once the problems of bit rate adaptation and cell delineation have been resolved, the digital flows have to be fed into the transmission system, whether framed or not.

For framed transmission systems, two cases of **mapping** a continuous flow of cells into the payload need to be considered:

- using a synchronous transmission medium (SDH);
- using a plesiochronous transmission medium (PDH).

Adaptation to synchronous transmission

In the case of an SDH synchronous transmission, the adaptation takes place at path level. ITU-T Recommendation I.432 specifies the mapping for STM-1 frames at 155.520 Mbit/s (as well as STM-4 frames at 622.080 Mbit/s). The STM-1 synchronous frame provides a capacity of 2430 bytes every 125 μs (155 520 kbit/s). These 2430 bytes are arranged in 270 columns and nine rows and are transmitted row by row. The first nine rows (81 bytes) do not carry data and are an overhead, used to delineate and manage the frame. Figure 2.10 shows the structure of an STM-1 synchronous frame.

The remaining 2349 bytes form a fourth-order virtual container, VC 4, which in turn is made up of a column (9 bytes) that contains the POH (path overhead) and the container itself, and provides a transmission capacity of 2340 bytes every 125 μs (149 760 kbit/s). The path overhead is used for management functions (parity check on the path, type of payload, continuity check on the path, and so on). Byte C2 (type of payload = ATM) is specific to the mapping of ATM cells into a C-4 container.

A whole number of 53-byte cells cannot be fitted into a payload of 2340 bytes. Furthermore, the flow of cells is mapped continuously into the C-4 container, so that some of them overlap onto the adjacent frame (Figure 2.11). In the case of this type of mapping, cell delineation is based on the HEC field.

Adaptation to plesiochronous transmission (Recommendation G.804)

In the case of a plesiochronous transmission, the cells can be grouped in a periodic frame, which also includes maintenance functions. The associated protocol, PLCP (physical layer convergence protocol), is derived from those defined by ETSI for IEEE standard 802.6, which is used for metropolitan area networks. It involves the 44 736-kbit/s PDH layer. Four bytes are added to each cell mapped into the transmission payload; the first two provide a delineation pattern, the third the order number of the cell in the 125-μs frame, and the fourth is reserved for management functions similar to those provided by the path overhead in the synchronous digital hierarchy. The overhead associated with this technique is quite large, and another method consists in simply mapping the cells into the payload of a PDH frame. For example, using this technique, 10 cells can

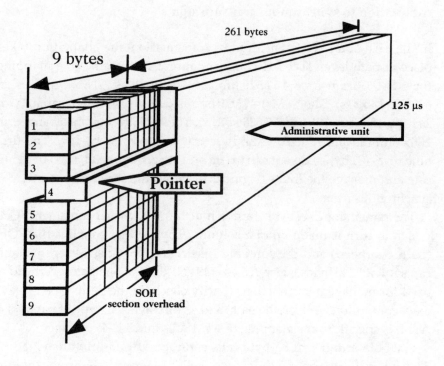

Figure 2.10 Structure of the STM-1 synchronous frame

be mapped into a 34 368-kbit/s frame, the HEC field being used for delineation.

Adaptation to an unframed transmission system

When the transmission system is unframed, other techniques must be used. The examples given below involve mapping cells mainly onto local area network systems and are based on proposals made by the ATM Forum:

- ATM transmission at 155.520 Mbit/s is achievable on an optical multimode fibre using 8B/10B coding. This type of transmission is possible over distances up to 2 km. Shorter distances (100 m) can be covered on STPs (shielded twisted pairs) with the same information coding. Another type of coding (NRZ,

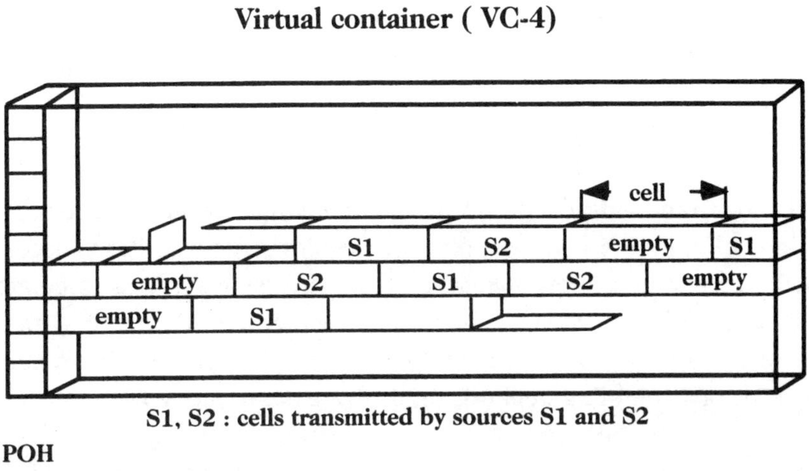
Virtual container (VC-4)
cell
S1
S2
empty
S1
empty
S2
S1
S2
empty
empty
S1
S1, S2 : cells transmitted by sources S1 and S2
POH

Figure 2.11 Adapting a flow of cells to the STM-1 synchronous frame

non-return to zero) allows use of UTPs (unshielded twisted pairs) over the same short distances (category 5, or 'data' quality).

- In the case of ATM transmission at 100 Mbit/s using 4B/5B coding and the physical media developed for FDDI networks, the start of each cell is identified by a symbol 'TT', whereas symbols 'JK' are used for stuffing between cells. To ensure proper synchronization, a symbol 'JK' is needed at least twice a second. This 100 Mbit/s interface is often called the 'TAXI (transparent asynchronous xmitter receiver) interface' after the integrated circuits developed for implementing the FDDI physical layer functions.
- Category 3 ('voice' quality) UTPs can be used at a rate of 51.84 Mbit/s using CAP 16 (carrierless amplitude modulation/phase modulation) code. The 16 symbols of this code carry 4-bit characters and thus reduce the modulation rate required to 12.96 Mbaud.
- Similarly, a data rate of 25.6 Mbit/s is achievable over category 3 UTPs, with 4B/5B coding, which leads to a modulation rate of 32 Mbaud. NRZI (non-return to zero inverted) is used to

represent data and the start of each cell is identified by a specific byte. To ensure 8-kHz synchronization, another byte is also available on this interface, and can be inserted in a cell.

2.3.5 Scrambling cells

Cells are scrambled in order to protect them against false headers, either accidental or intentional. Scrambling also allows the transitions necessary for the correct operation of cell-based transmission systems to be generated. The method consists in adding the sequence of data items, modulo 2, to a pseudo-random sequence produced by a generator polynomial. The type of scrambling depends on the transmission environment.

- In the case of a cell-based transmission system, all the fields in the cell are scrambled except the HEC field. During the cell delineation acquisition phase, pseudo-random sequence synchronization information must be transmitted to the receiver. This information is provided in 2 bits of the HEC field in several cells. During this phase, cell delineation is carried out on only 6 bits, which explains why more confirmations are needed (delta = 8) than in synchronous transmission. The polynomial for this type of scrambling is $x^{31} + x^{28} + 1$.
- When a synchronous physical layer is used, only the payload is scrambled. Although this system is simpler, it has the disadvantage of doubling transmission errors (to be more precise, the error rate is multiplied by the number of terms of the polynomial used). The polynomial that has been chosen is $x^{43} + 1$.

■ 2.4 Functions of the AAL

2.4.1 Overview

The ATM layer, described above, provides a single, high-speed switching service for all the flows generated by applications with very varied profiles. These flows are switched and multiplexed by common mechanisms, and only multiple queues, located before these mechanisms, can provide differentiated handling. The service provided by the ATM layer can be summarized as follows.

- Cell relay operates in connected mode and therefore preserves the sequence order of the cells transmitted.
- The service operates independently of the traffic source's clock. However, this advantage implies that there is no explicit information about the source's clock in the flow received. Furthermore, this asynchronism, added to the presence of queues in the network, introduces variable propagation delays which cause cell jitter of about 0.1 ms.
- It does not provide flow control. If necessary, this must be added in the higher layers (user applications).
- It is totally transparent to the cells' payloads. It does not change their contents, but provides no means of checking their integrity. Further, if a transmission error concerning the header is wrongly corrected (when there are multiple errors), there is a risk of imitating a valid logical identifier, and therefore inserting the cell on that logical channel. Conversely, if the HEC correction capacity is exceeded, cells can be lost, which can also happen when the network queues are congested.

The AAL layer is much more strongly linked to the applications: it refines the quality of service provided by the ATM layer according to the requirements of the user service. It implements end-to-end protocols that are transparent to the ATM layer.

In particular, as the information being transferred has no particular reason for being compatible with the length of the ATM cell (48 bytes), the information has to be segmented or grouped on transmission and the cells' contents reassembled or divided on reception.

Different services would therefore require specialized adaptation layers; however, to avoid too great a dispersal of efforts, classes of service have been grouped around three main components, which characterize all traffic flows:

- its bit rate, which may be constant or variable;
- its connection mode, which may be connection oriented or connectionless;
- its requirements from an isochronous point of view, which may impose a strict relationship between the source's clock and that of the receiver, or no relationship whatsoever.

Four types of adaptation derived from combinations of the characteristics mentioned above were first defined: AAL types 1, 2, 3 and 4. Subsequently, AAL types 3 and 4 were combined into one, called AAL type 3/4, and a new AAL type 5 appeared, due to pressure from the computer world.

This historical reminder illustrates the fact that the list of adaptation mechanisms is not necessarily closed, given their close relationship with users' applications. At present the standardized adaptation mechanisms are (Figure 2.12):

- AAL type 1, for constant bit rate information which requires a strict relationship between the transmission and reception clocks (for example, voice circuit emulation);
- AAL type 2, for variable bit rate information, which also requires a strict relationship between the transmission and reception clocks (for example, variable bit rate video);
- AAL type 3/4, for data transmissions in connection-oriented or connectionless mode;
- AAL type 5, which can be seen as a simplified version of AAL type 3/4 but with similar capabilities.

These adaptation layers are structured into two sublayers.

- The segmentation and reassembly (SAR) sublayer is responsible for changing the format between the user data units and the cell payloads. The AAL fields corresponding to this sublayer, which is relatively independent of the user service, are present in every cell. This function enables lost or duplicated cells to be detected, because they are numbered; nevertheless, recovery itself is the province of the convergence sublayer. Finally, the SAR sublayer provides for the padding of incomplete cells.
- The convergence sublayer (CS) carries out more specifically user service functions. The AAL fields related to these functions are only present once per user data unit. The convergence sublayer is responsible for error handling if needed: to do so, it implements protocols for retransmitting erroneous data or it protects data, allowing the receiver to correct those errors. This last technique, forward error correction (FEC), is used in partic-

Temporal relationship	with		without	
Bit rate	continuous		variable	
Connection	with			without
	AAL-1	AAL-2	AAL-3/4 or AAL-5	AAL-3/4

Figure 2.12 The types of ATM adaptation

ular for real-time applications. The CS can also provide end-to-end synchronization.

2.4.2 AAL type 1 adaptation function

The AAL type 1 adaptation function is used by applications with strong isochronous constraints and a constant bit rate, such as:

- voice signals;
- high-quality audio signals;
- video signals;
- data circuit emulation.

Its role is to allow the transmitted information's clock pulse to be recovered, to compensate for the differences in propagation time caused by the network and to manage the loss or accidental insertion of cells. It also allows for the use of block-structured data and provides means for handling errors.

The fields related to these functions occupy a byte of the payload, leaving 47 bytes available for the information (Figure 2.13). They include a sequence number (SN), used to detect missing or accidentally inserted cells, and a sequence number protection (SNP). The SNP field is divided into two:

- a 3-bit cyclic redundancy code (CRC) for correcting single errors;
- a parity bit for detecting double errors.

The SN field is also divided into two.

- The first bit, convergence sublayer information (CSI), may contain a residual time stamp (RTS), which is used for setting the receiver's clock. It may also be used to delineate the data blocks.
- The next 3 bits contain the counter for numbering (modulo 8) the cells.

In certain continuous flow applications, such as data circuit emulation or high-quality audio signals, it is necessary to reconstitute the information received precisely and therefore to absorb cell jitter totally. To do this, the receiver must store the information received for a period of time that is at least equal to the propagation time, plus the maximum value of the jitter (Figure 2.14).

An asynchronous type transfer, such as ATM, does not allow the source's clock to be controlled by the network clock, so the receiver must restore the source's clock. The cell flow may therefore contain a time stamp (RTS) given by a reference clock. This technique requires the use of a common reference clock derived from the underlying transport network (SDH, for example). This 4-bit time stamp is carried by the CSI bit of one cell in two belonging to a group of eight consecutive cells (odd-numbered cells).

In the case of voice or video signals, it is not usually necessary to use the above method, called SRTS (synchronous residual time stamp), and the receiver can recover the source's clock approximately from the rhythm at which its memory buffers are filled. The receiver's average incoming filling rate is used to set an initial cell restitution delay. If the filling rate increases, the cell restitution rate accelerates, and conversely if it decreases (Figure 2.14).

The AAL type 1 adaptation function also allows block-structured data to be transferred (for example, to support a data circuit at $n \cdot 64$ kbit/s). A pointer, which indicates the location of the next block, makes it possible to delineate the block. This pointer occupies the first byte in the payload (then limited to 46 data bytes) and its presence is indicated by the CSI bit.

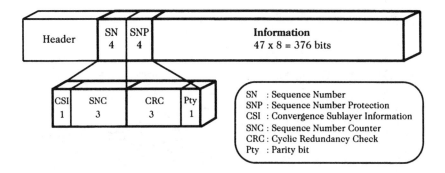

Figure 2.13 AAL type I

To ensure compatibility with the SRTS method, described above, the pointer can only be present in even cells.

To compensate for the delay introduced by the assembly of bytes into cells, especially in case of a digitized telephone signal (p. 9), AAL type 1 allows the use of partially filled cells. The number of meaningful bytes per cell (fewer than 46 or 47) is then an optional parameter supplied at ATM connection set-up time.

The AAL type 1 convergence sublayer can also take responsibility for error handling: single or multiple errors in the transmission of the cells' payloads, but also loss or accidental insertion of whole cells, detected by the SAR sublayer. The handling mode and its performance depend on the type of user service: thus, in the case of data circuit emulation, and on condition that the error rate and the cell loss rate are very small, errors can be handled by end-to-end protocols, without AAL layer intervention; similarly, no handling is needed for the telephone service.

Simply detecting a lost cell, with a view to retransmission, is not suitable for real-time applications, because it would involve an unlimited response time. However, it can contribute to the masking of errors by a received information interpolation mechanism. This masking is particularly effective if data has been interleaved on transmission, because this allows the effects of cell loss to be diluted (see the description below).

If error masking is not sufficient to satisfy the user's service constraints (for example, in the case of video signals or high-quality audio signals), the original information has to be reconstituted using an FEC-

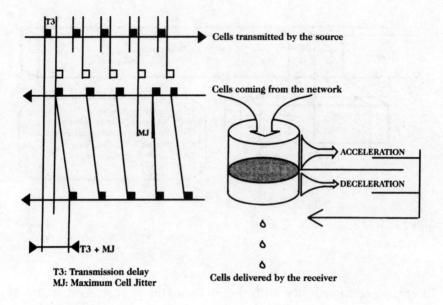

Cells transmitted by the source

Cells coming from the network

ACCELERATION

DECELERATION

T3 + MJ

T3: Transmission delay
MJ: Maximum Cell Jitter

Cells delivered by the receiver

Figure 2.14 Receiver's clock controlled by the transmitter

type correction technique, which may be combined with an interleaving mechanism.

The principle of **byte interleaving** is based on a very simple idea: if the information is transmitted in cells as and when it is generated, the loss of one cell on reception affects 47 consecutive bytes. On the other hand, the information to be transmitted can be stored temporarily in blocks of 47 times p bytes, after which p cells are made up in the following way: the first contains the bytes numbered 1, $p + 1$, $2p + 1$..., the second the bytes numbered 2, $p + 2$, $2p + 2$... and so on until the cell numbered p is reached. If a cell is lost (the number of that cell is known by the receiver because it detects a sequence break in the SAR sublayer), the reconstituted received flow is only affected by one byte every p bytes, which makes interpolation much easier. We should note that the first cell in p is identified by the CSI bit, which makes this method incompatible with the transfer of block-structured data.

If it is necessary to reconstitute the original set of bytes exactly, a distributed correction technique is added to the interleaving mechanism:

instead of storing $47 \cdot p$ bytes directly, q bytes are added to each 'line' of p bytes generated, giving a total of $47 \cdot (p + q)$ bytes. These q bytes are redundant information, calculated from the corresponding p bytes. When the storage memory is read, $p + q$ cells are transmitted, of which q are redundant (Figure 2.15). In practice, the ITU-T recommends the use of a Reed Solomon code RS(128,124), that is $p = 124$ and $q = 4$. If no cells are lost, this code enables up to 2 bytes to be corrected per 'line' of 128 bytes, after reverse interleaving. If the receiver's SAR sublayer detects a lost cell, it provides a dummy cell as a replacement, with an error indication: the correction code can then correct up to four missing cells.

Naturally, such a technique has the disadvantage of introducing a delay of $p + q$ cells, both on transmission and reception; in the above case, a delay equivalent to 256 cells. The lower the bit rate, the greater the problems caused by this delay: in a 384-kbit/s television telephony application, this mechanism leads to a delay of 250 ms.

Other methods can be imagined. As an example, it is possible to store an amount of information corresponding to 16 cells in the form of a matrix of eight lines and 94 columns, of which six are redundant. By using a diagonal interleaving technique and a Reed Solomon code RS (94, 88), this method allows one lost cell to be corrected per block of 16 cells, while limiting the additional delay.

2.4.3 AAL type 2 adaptation function

The role of this adaptation function is similar to that of the AAL type 1 function, as far as the clock recovery, jitter compensation and cell loss and insertion management functions are concerned. However, the fields corresponding to these functions have to be adapted to the transmission of variable-length data units. The format of these fields is still under study.

2.4.4 AAL type 3/4 adaptation function

The AAL type 3/4 adaptation function operates either in connection-oriented mode or in connectionless mode, in which the data units (datagrams) are routed independently from each other. The flow carried in connection-oriented mode can be either assured or not by the network.

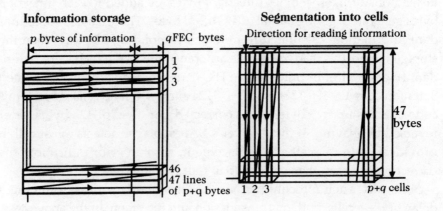

Figure 2.15 The byte interleaving technique

- In **assured** mode, the AAL type 3/4 layer implements flow control and missing or faulty unit retransmission functions.
- In **non-assured** mode, these functions must be provided by the higher layers.

These adaptation functions accept data units with a maximum length of 65 535 bytes and provide two priority levels: normal priority and high priority.

As Figure 2.16 shows, the CS is made up of two parts called the CPCS (common part convergence sublayer) and the SSCS (service-specific convergence sublayer). The functions related to the assured and non-assured modes of the connection-oriented service are carried out in this latter part.

The functions provided by the CPCS are the following:

- service data unit (CPCS-SDU) delineation;
- error detection (optionally, the faulty CPCS-SDUs can be transmitted to the higher layers with an error indicator, if not they are discarded);
- receiver information about memory needed to receive the CPCS-SDU;

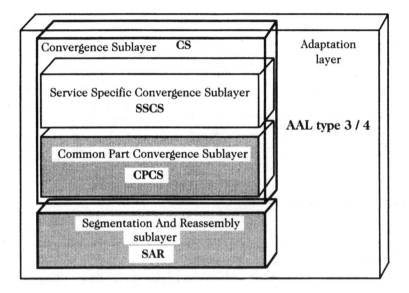

Figure 2.16 AAL type 3/4

- transmission of an 'abort' message.

The CPCS functions provide support for a connectionless service as well as a connection-oriented service.

The significant fields of a CPCS-PDU are (Figure 2.17):

- a common part indicator (CPI) field which indicates how the following fields should be interpreted;
- CPCS-SDU beginning and end (**Btag, Etag**) indicators, used to prevent accidental concatenation of two CPCS-SDU, resulting from the loss of the cells carrying the end of the first data unit and the start of the second;
- a CPCS-SDU size indicator (**BASize**), which tells the receiver how much buffer memory to reserve (so that the receiver does not have to systematically reserve the maximum size of buffer memory, 64kb);
- padding to align the CPCS-SDU on a 32-bit boundary (AL, alignment);

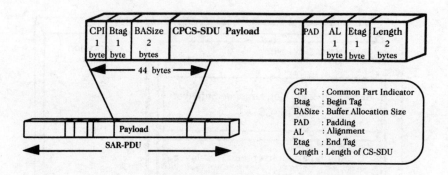

Figure 2.17 CPCS-PDU structure

- a final CPCS-SDU size (**Length**) indicator, which gives the exact length of the payload so that the padding can be eliminated.

The SAR sublayer ensures the integrity of the cell's payload and provides grouping or ungrouping and segmentation or reassembly functions: units (CS-PDU) smaller than a cell's payload are grouped, whereas larger units are segmented. The fields that provide for these management functions occupy 4 bytes in the cell, thus reducing the payload to 44 bytes (Figure 2.18). We should note that this payload size, a multiple of 4 bytes, is adapted for use by 32-bit processors.

The SAR sublayer fields include:

- a segment type (ST) indicator: start, middle or end of message, or message composed of a single segment;
- a modulo 16 sequence number (SN) for detecting missing or inserted cells;
- a priority indicator, used to transmit high-priority SAR-PDU before normal priority ones;
- a multiplexing identification (MID) indicator, used for identifying cells belonging to different data flows (maximum 512) multiplexed on the same virtual connection;
- a length indicator (LI), which gives the number of bytes (from 1 to 44) used in the cell;
- a 10-bit CRC, which enables errors to be detected in the cell's payload.

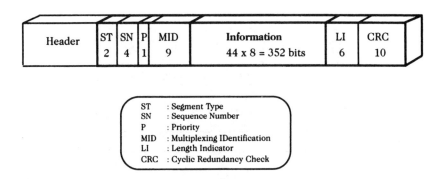

Figure 2.18 SAR-PDU structure

A special coding of a SAR-PDU is used to transmit the abort message: the segment type indicates 'end of message' and the payload is set to zero, as is the LI field.

If a cell relay network is to provide a connectionless service then it must provide additional functions in order to ensure the routing of the datagrams (CLSF, connectionless service function). These functions are defined in a higher layer than AAL type 3/4, which is used in non-assured mode, called CLNAP (connectionless network access protocol) (Figure 2.19).

This set of protocols is very similar to that defined in the DQDB standard: addresses based on the E.164 numbering plan, address validation, address groups supported and so on.

The main fields of the CLNAP layer are the following:

- the datagram's source and destination addresses, necessary for routing;
- a higher layer protocol identifier (HLPI) field;
- a QoS indicator;
- an optional CRC, for error detection.

2.4.5 AAL type 5 adaptation function

This function can be seen as a simplified version of layer AAL 3/4: just like that layer, its aim is to transfer data flows made up of variable-length

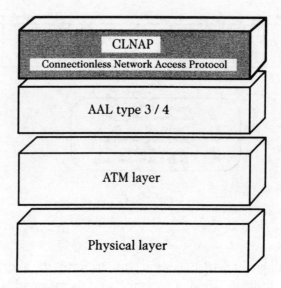

Figure 2.19 Connectionless service

units with a maximum length of 65 535 bytes in connection-oriented mode.

The data unit sent to AAL type 5 by the user is padded so that it makes a unit which can be divided into an integer number of cells: the length of the resulting unit is then a multiple of 48 bytes, and the length of the padding is between 0 and 47 bytes. The last cell contains 8 bytes dedicated to three different functions:

- a length indicator (16 bits), which the receiver uses to determine the data payload;
- a 32-bit CRC, used to detect errors in the data transferred;
- 16 bits reserved for future use.

Because all the payload's 48 bytes are occupied by user data, the AAL type 5 function cannot provide any explicit start and end of data unit indicators. The data unit's last cell marker is provided by the last bit of the PTI field (type of SDU = 1) in the ATM header (p. 20). Therefore, the first cell of a data unit is implicitly identified as that which follows the last cell of the previous unit, which is explicitly determined (Figure 2.20).

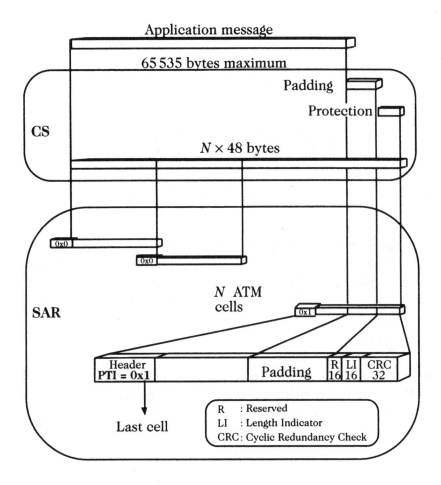

Figure 2.20 AAL type 5 operation

The AAL type 5 function is adapted to a hardware implementation of the CPCS. Its advantage is that it uses the whole of the cells' payloads and implements an effective protection of the data unit using a 32-bit CRC. However, it does not allow cell-by-cell error detection or the multiplexing of several flows.

■ 2.5 Maintenance flows

An ATM network must be able to measure the quality of service that it is offering to its users and detect any degradation. It needs to have ways of locating any faulty components so that it can undertake any necessary reconfiguration. Pinpointing the exact location is particularly valuable when the network is complex.

Just as the ATM layer handles connections made up of virtual paths and virtual channels, the underlying transmission system is structured into several levels: transmission media, regeneration sections, multiplexing sections and transmission paths. Furthermore, an ATM network may be divided into several subnetworks (or segments), which may be administered separately.

The maintenance flows are defined in ITU-T Recommendation I.610. They are responsible for the following functions:

- performance management, consisting of using a parity check (BIP, bit interleaved parity) and gathering results (FEBE, far end block error) to evaluate the error rate;
- fault management, using continuity tests and mechanisms for signalling events (AIS, alarm indication signal) or returning backwards fault indications (FERF, far end receive failure).

There are five maintenance flows, as shown in Figure 2.21. Flows F1, F2 and F3 are carried by channels provided by the physical layer, depending on the type of support (framed or continuous flow of cells); flows F4 and F5 use virtual connections (paths or channels) provided by the ATM layer.

2.5.1 Physical layer maintenance flows

Maintenance flows F1, F2 and F3, responsible for monitoring the regeneration section, the multiplexing section (also called the digital section) and the transmission path respectively, mainly use means specific to the transmission system.

In the case of an SDH physical layer, the maintenance flows use the synchronous hierarchy's section (SOH) and path (POH) overheads (Figure 2.10). Performance measurement is carried out on blocks of bytes whose size is exactly equal to the payload of the virtual containers (2340 bytes for a 155.520-Mbit/s STM-1 frame, 9360 bytes for a 622.080-Mbit/s STM-4 frame). Therefore, the parity check is applied to all the bytes in the cells (headers included) transported by the containers. In the case of flows F1 and F3, it is carried out byte by byte (BIP-8), whereas in

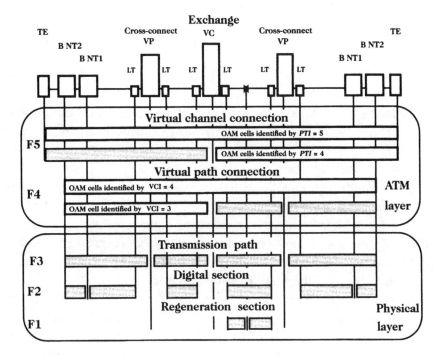

Figure 2.21 ATM maintenance flows

the case of flow F2 it is applied to 3-byte words (BIP-24) at 155.520 Mbit/s or 6-byte words (BIP-96) at 622.080 Mbit/s.

Similarly, the PDH physical layer uses certain binary elements of the 34.368- and 139.264-Mbit/s systems' overheads for the maintenance flows. The parity check (BIP-8) is again applied to the whole of the payload, that is 530 or 2160 bytes.

The physical layer of a cell-based system does not provide a priori any specific means of communication for the maintenance flows. In this case, special maintenance cells (OAM, operation, administration and maintenance) are regularly inserted in the cell flow. They are identified by a specific header, which also indicates whether it is an F1-type flow or an F3-type flow (note that a cell-based system does not have a multiplexing section). Performance measurement is carried out on a fixed number of cells. The OAM cells may contain parity check information (BIP-8), results (number of parity errors) or AIS and FERF indications.

Their contents are protected by a 10-bit CRC (polynomial $x^{10} + x^9 + x^5 + x^4 + x + 1$).

2.5.2 ATM layer maintenance flows

Whereas the physical layer flows can only be accessed by the network operator, the F4 and F5 maintenance flows can be used by the user. Generally, they are only activated on request. They are end-to-end flows, but there are also segment flows (subnetwork flows).

A virtual path is checked (flow F4) by sending OAM cells on a reserved virtual channel (VCI = 4 for an end-to-end F4 flow, VCI = 3 for a subnetwork F4 flow). Maintenance flows involving a given virtual channel (flow F5) take the same path as the operational cells: they are distinguished by a special coding of the PTI field in their headers (PTI = 5 if the F5 flow is end to end, PTI = 4 if it applies to a subnetwork).

The F4 and F5 flows use the same mechanism for measuring performance. It is implemented on nominal sized blocks (N = 128, 256, 512 or 1024 cells). The OAM parity check (BIP-16) cell is only inserted after N cells when there is no activity, so as not to cause jitter in the operational flow. Insertion is forced if there has been no activity for $3N/2$ cells; the next insertion remains fixed at $2N$ and the protected block then has a reduced size of $N/2$ cells.

To check that a connection is still active, test cells can be sent when no working cell has been transmitted during a given period and no fault has been signalled.

■ 2.6 Cell relay performance

The parameters that affect cell relay performance are:

- loss of cells;
- transfer delay.

2.6.1 Loss of cells

There are two major causes of cell loss: header errors and buffer overflows. Even though the header is protected against errors, some are neither corrected nor detected, which leads to routing errors. Furthermore, error handling by the HEC leads to cells whose headers contain uncorrectable errors being discarded. Cell relay is therefore subject to a cell loss rate and an incorrect routing rate.

The sizes of the buffer are not infinite, and there is a probability of cell loss as a result of overflow. This loss depends on the number of flows multiplexed on the same route, on the sizes of the buffer located on that route and on the nature of the flows multiplexed. The buffer sizes and the number of flows multiplexed are determined according to the acceptable cell loss target. Table 2.2 shows targets for different information flows.

The numbers indicated in this table are widely divergent. These values must be maintained to guarantee a good quality of service for each flow. One way of guaranteeing a good quality of service for each information flow is to provide a large number of service classes. However, any increase in the number of service classes also increases the complexity of the network.

The use of the cell loss priority bit helps in achieving a cell loss target, for a given class of service, by indicating cells carrying less important information that can be discarded (p. 24). This technique is feasible for video flows, but it is more difficult to implement for data transmissions, where all the bits are a priori equally important.

2.6.2 Transfer delay

For a given information flow, the global delay affecting the cells depends on three main factors:

- T1, the coding and decoding time;
- T2, the time required for segmentation and reassembly;
- T3, the time taken to transfer cells across the network.

Information coding and decoding

T1 depends on the type of coding used: a G.711-type coding (PCM at 64 kbit/s) only adds a few milliseconds, whereas a more sophisticated technique produces less information for transmission but takes more coding and decoding time (several tens of milliseconds).

Segmentation and reassembly

T2, the time required for segmentation and reassembly, can be broken down into two factors: the transmitter segmentation delay (**T21**) and the delay (**T22**) introduced in the receiver to compensate for the variations in cell transfer time.

Segmentation of M bytes of information into cells causes a delay, T21, which depends on the bit rate D (expressed in bits per second), and can be approximately evaluated by $8M/D$. This delay decreases as the bit rate increases.

Cells inserted at a regular rhythm by an information source are not delivered regularly by the network (p. 42). Asynchronous transfer mode introduces variations in the transfer delay. The receiver must compensate for them, and therefore add a delivery delay that will allow it to absorb the largest differences.

Transfer across the network

T3, the time taken to transfer cells across the network, is the sum of the propagation delay of the transmission media (T31) and the transit time (T32) spent in the switching nodes.

T31, the propagation delay, is a function of the distance and the number and nature of the physical media used between the source and the destination. Transmission on fibre optics introduces a delay of about 5 ms per 1000 km. The longest terrestrial distance produces a delay not longer than 50 ms, compared with 300 ms for a satellite link.

T32 includes the time the cells have to wait in the buffer and the time taken to insert them into the physical transmission supports. This time is a direct function of the bit rate used (about 3 μs per cell at 155.520 Mbit/s). The time the cells have to wait in the buffer depends on the dimensioning of the nodes: an average filling level of 100 cells causes an average delay of 300 μs per node (100×3 μs). The global delay then depends on the number of nodes the cell passes through.

Global delay target

The information flow that generates the most restrictive delay target is voice: let T0 be that target's value. Two of the delay components

Table 2.2 Cell loss rate targets

	Format	Target
Telephone-quality voice	ITU-T G.711	$<10^{-3}$
	PCM (64 kbit/s)	
High-quality voice	ITU-T G.727	$<10^{-5}$
	SB-ADPCM (64 kbit/s)	
Standard-quality television	Signal compression (10 Mbit/s on average)	$<10^{-9}$
High-definition television	Signal compression (100 Mbit/s on average)	$<10^{-10}$
Data transmission	HDLC (64 kbit/s to 100 Mbit/s)	$<10^{-6}$

described above are variable and influence the dimensioning of the network (the number and the capacity of links and switching nodes):

- T22, the transfer delay variation compensation time;
- T32, the time spent waiting in the switching nodes' buffers.

The other components (T1, T21 and T31) are more or less fixed and known for any given environment. T22 and T32 must therefore be selected so that the following is true: T22 + T32 < T0 – T1 – T21 – T31.

3 | ATM switching units

▦ 3.1 Introduction

A complex network is made up of interconnected switching units. The role of a switching unit is to set up a connection between an input port and an output port, according to the routing information. Before describing the specific constraints of ATM, we will summarize the principles and limitations of conventional switching modes.

3.1.1 Circuit switching

The technique known as 'space division switching' consists in physically linking an input port of the switching unit to one of its output ports for the duration of a communication (for example, a telephone call). This technique introduces a constant and very short delay, because the switching unit does not store the information.

Time division synchronous switching uses time slots which are assigned to the channels being switched. It operates with physical supports that are time division multiplexed according to a fixed-length frame structure. The information corresponding to an input multiplexer's time slot is stored temporarily and then delivered at regular intervals, in an equivalent frame but in a different time slot, to one or more output multiplexers chosen by the switching unit. The correspondence between the time slots of the multiplexers, which switch between input and output channels, is independent of the use made of the channels and the bit rates depend not on the sources of information, but solely on the characteristics of the multiplexing system being used (bit rate, frame structure).

3.1.2 Packet switching

This is a case of asynchronous time division switching: the packets, made up of data blocks accompanied by a pointer contained in the header, are received on the switching unit's input links at a speed that depends solely on the source. Each packet is stored and then delivered to the output link determined by the routing information contained in a table. The entry accessed in the table depends on the pointer's value. The storage time, and therefore the time the packet is delayed, is variable because of the statistical sharing of resources.

A feature of a datagram service is the fact that the packets are independently switched according to their explicit destination addresses, and that no previous marking is necessary. By contrast, in the case of a 'logical channel' type service, where the sequence of packets must be maintained, the technique used consists in making the packets for a given connection follow the same path, identified by a series of pointers. In practice, in a complex network, there may be several possible paths between a given input point and a given output point.

The switching function is usually carried out by software. Other functions, such as flow control and retransmission on error, may be undertaken by the same processor or may be left to the accesses. Normal performance is several thousand packets switched per second, with a composite global speed of several Mbit/s; the delay is in the order of 10 or 100 ms.

The required characteristics of an ATM switching unit are completely different:

- very high access speeds mean that there will be a global bit rate of several Gbit/s;
- several million cells are switched per second;

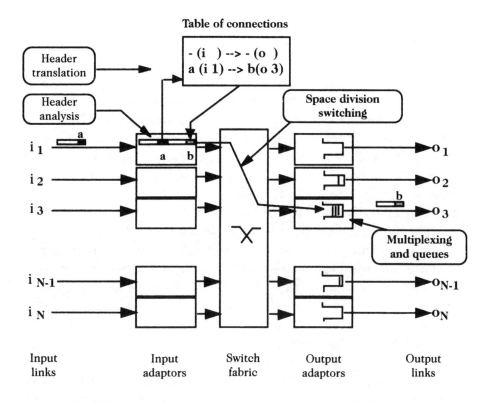

Figure 3.1   ATM switching unit functions

- there must be a stable and very small delay (less than 1 ms), to provide circuit emulation;
- there must be a low cell loss rate.

This is only possible using a hardware switch fabric which is highly parallel. This switch fabric may be made up of several identical switching components, which may be organized into a multilevel structure.

▨ 3.2 ATM switching unit functions

As well as analysing and modifying the header (new VPI/VCI values), an ATM switching unit provides two main functions, described below (Figure 3.1):

- **routing** cells to the appropriate output ports;
- temporarily **storing** cells.

The switching unit must also manage several parallel cell flows differentiated by **priority** levels and provide preferential handling for high-priority cells (for example by implementing queues for each priority level).

Finally, certain ATM services require the **broadcasting** of cells coming from the same source: broadcast, to all destinations, or **multicast**, to a predetermined set of destination accesses. Of course, the source itself could provide distinct copies of each of the cells to be broadcast, but these copies would be routed as so many independent flows, which would lead to a waste of bandwidth and would mean that the source would have to know the complete list of destination addresses.

A more efficient method is to shift the point where the cells are duplicated as far forward as possible: using a particular item of information (broadcast address), an ATM switching unit should be able to replicate the same cell to several output ports.

As shown diagrammatically in Figure 3.1, the header is usually handled by the switching unit's input adapters, which are responsible for recovering the flow of valid cells from the input links. The routing is carried out by the switching fabric. The temporary storage of cells usually takes place in the output and/or input adapters, but may also be centralized in the switching fabric.

3.2.1 Routing cells

Generally, the connection between an input port and an output port, which determines the route through the switching fabric, must be previously known and stored in the table of connections. This information may be held as a marker which establishes a specific path for transferring the cells related to a given connection (**indirect routing**), or as a label which, when added to the cells to be transferred, allows them to direct themselves to the appropriate output port (**self-routing**).

As the ATM service is connection oriented, the natural routing mode is indirect: each cell's header contains a pointer (VPI/VCI) whose value identifies the connection and which only has a local meaning (p. 21). The path corresponding to that connection must be explicitly written to every switching component before any information is transferred. Cells are then

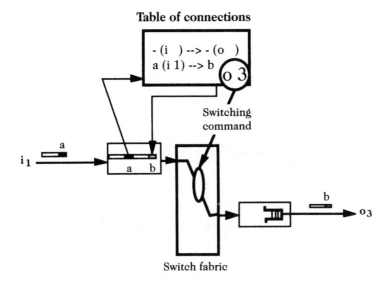

Figure 3.2 Indirect routing in an ATM switching unit

routed to the appropriate output port by consulting the table: for each VPI/VCI value there is a corresponding output port and a new pointer value (Figure 3.2). Potentially, the table can be very large, because each connection is identified by 24 or 28 bits in the cell's header.

One way of implementing this routing, at the cost of an overhead, consists in using self-routing: at the input to each switching unit (especially if it is made up of several switching components), an additional routing label is added to each cell. It describes the physical route the cells must take and is in the form of a list of identifiers of the switching components they must traverse and the output ports they must use (Figure 3.3).

All the cells related to a given connection follow the same route and are delivered to the receiver in sequence. The switching components do not mark them, because the route is written explicitly in each label. The information that has been used is removed from the label's contents (sometimes called a 'consumable label') as the cell makes its way through the switching unit (instead of deleting part of the label, a pointer that indicates that the remaining useful part can be modified).

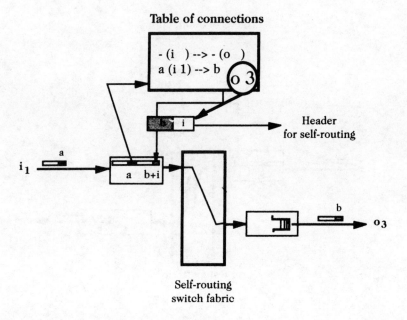

Table of connections

- (i) --> - (o)
a (i 1) --> b

o 3

i

Header
for self-routing

i_1

a

a   b+i

b

o_3

Self-routing
switch fabric

Figure 3.3 Self-routing in an ATM switching unit

3.2.2 Temporary storage of cells

In certain switch fabrics there is a risk of internal blocking because it is not always possible to establish a path between an input port and an available output port (in particular, this is the case for Banyan-type switching units described on p. 69). It is therefore natural to think that, to avoid unacceptable loss rates, storage devices (queues) are essential at the inputs or inside the switching unit.

However, even if a non-blocking switch fabric is used, cells still have to be stored temporarily to resolve **output contention**. In practice, owing to the statistical nature of the incoming traffic, several cells received on different input ports can be in competition for simultaneous access to the same output port.

There are two classic approaches to the location of queues: **input queuing** or **output queuing** in the switch. There are others, but for the most part they can be seen as variants or combinations of these two. The

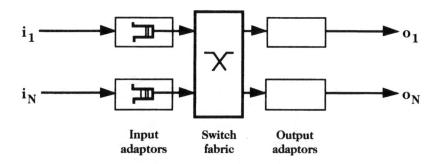

Figure 3.4 Input queuing of cells

following paragraphs describe these methods, using the following hypotheses:

- the switching unit involved is of the $N \times N$ type (N input ports and N output ports operating at the same speed);
- the input traffic flows are independent and statistically identical; they are uniform, and each cell has a probability equal to $1/N$ of having a given output point as its destination.

Input queuing

A FIFO (first in first out) queue is associated with each input port. A contention is detected if j cells ($j \leq N$) located at the head of j queues are destined for the same output. This approach is natural enough in that it detects the contention before the switching fabric is reached and only supplies cells that can reach their destination (Figure 3.4). However, all the cells further back in the $j-1$ queues that are not serviced are also blocked, even if they are destined for ports which are free at the time.

This head of line (HOL) blocking effect limits the performance of input queuing. It can be shown that, if N is large, the load offered by the switch fabric cannot exceed a value equal to $2-\sqrt{2}$, that is about 0.58, whatever the queue arbitration algorithm (e.g. random, cyclical), if it is equitable (note that for $N = 4$, the maximum load is already limited to 0.65).

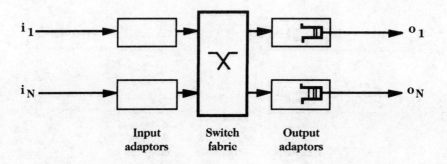

Figure 3.5 Output queuing (unlimited size queues)

The input queuing method is poorly adapted to broadcast functions because there are no output queues. Furthermore, it is very susceptible to non-uniform input traffic. On the other hand, it has the advantage of being very simple and does not require a switch fabric operating speed higher than the access speed.

Better performance can be achieved at the cost of greater complexity, for example by increasing the internal speed of the switch fabric. Another improvement consists in sorting the cells on entry according to their destination, which means managing N queues per input port (one for each output port).

Output queuing

In this approach, a FIFO queue is associated with each output port (Figure 3.5). All the cells presented to the input ports at any given time traverse the switch fabric at the same time and are then stored. As they may all have the same destination, the queue associated with the output port involved must be capable of storing N cells.

Fast conventional type queues can be used if the internal speed of the switch fabric is N times greater than the speed of the ports. Conversely, a high degree of parallelism will allow storage in multiport queues without requiring a switch fabric operating speed higher than the access speed. In both these cases, if the queues are of unlimited size, the use of the switch fabric is optimized, and no input queuing is required because there can be

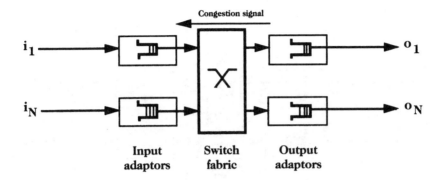

Figure 3.6 Output queuing (limited size queues)

no blocking. This technique is suitable for broadcast functions and is not affected very much by non-uniform input traffic. However, implementing it is extremely complex.

In terms of the bit rate provided or the delay, it can be proved that performance is practically stable once the output queues are large enough to contain 10 cells, as we find that statistically the bursts destined for a given port are of limited duration. However, the loss rate due to overflow is not negligible, which means that supplementary input queues need to be installed (Figure 3.6); the size of these queues depends on the real characteristics of the traffic and the admissible loss rate.

A congestion signal (**backpressure**) is used to keep the cells queued on input when the target output queues are full. This backwards storing of excess cells is subject to the **HOL blocking** effect.

An interesting improvement consists in considering all the output queues, which are all of limited size, as a single set of buffers which are allocated dynamically (Figure 3.7).

Using a centralized storage such as this reduces the amount of storage necessary for the same level of performance by a factor of 3 or 4, compared with the technique of dedicated output queues. An overloaded output port can temporarily use several buffers, which statistically reduces the use of the supplementary input buffers. This statistical multiplexing effect becomes more marked as the number of ports, N, increases.

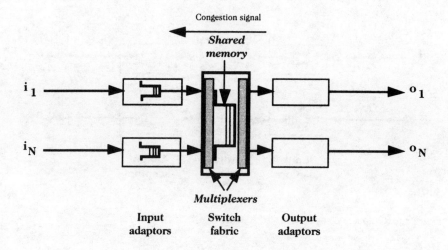

Figure 3.7 Centralized cell queuing

▧ 3.3 Types of switching fabric

ATM switching units can be grouped into two categories, according to their architectures: shared resource switching units and space division switching units.

3.3.1 Shared resource switching units

These work on the principle of multiplexing all the input flows towards a very high-capacity shared resource.

Some switching units, organized around a **shared memory**, benefit from the advantages of the centralized storage described above. The management of this centralized memory is complex, and its bandwidth is very broad, which means a high degree of parallelism has to be implemented in order to overcome technological constraints.

Others use a **shared medium** to connect the input ports to the queues associated with the output ports. This medium is usually a bus or a ring that transfers several bits in parallel.

3.3.2 Space division switching units

Switching units in this category are characterized by the coexistence of simultaneous paths between the input and output ports.

In the case of **crossbar**-type switching units, originally developed for circuit switching, a switching fabric with N inputs and N outputs has N^2 crossing points and there is no internal blocking: it is always possible to set up a path between an input port and an unoccupied output port, and simultaneous paths can be set up between unrelated pairs of ports. Output contention is resolved by input queuing or by queuing in the crossing points themselves. This latter technique is similar to distributed output queuing with, however, the disadvantage that the global storage capacity cannot be shared dynamically.

Banyan-type switching units have the advantage of only requiring $(N/2) \cdot \log_2 N$ switching components to form a matrix of N inputs and N outputs. For example, an 8×8 matrix requires 12 switching components organized into three stages of four components each. These are of the 2×2 type, which can connect each input to one of two outputs, depending on a cell destination address bit (self-routing). However, switch fabrics of this type can become blocked internally: there is only one path between a given input and a given output and there may be contentions for the use of an internal link.

Conventional input queuing and queuing inside the switching components are possible, but it can also be demonstrated that a Banyan network will not block if its inputs are ordered in relation to its outputs, in so far as there is no more than one cell per output port. This sort function can be carried out by a supplementary switching network (**Batcher** network) located in front of the Banyan network. One technique for the resolution of output contention, when several cells are destined for the same port, consists in only letting one of them through and sending the others back to the input to be sorted again.

4 Broadband ISDN

4.1 Overview

The aim of the broadband integrated services digital network (ISDN) is to carry all types of information (voice, sound, video, text, images and data) on a single network. Installing a limited number of interfaces is one of the essential aims of this multiservice network. In this context, ATM technology is particularly suitable for the multiplexing and switching functions. The services should be available on virtual connections which are set up permanently or on request, that is call by call. Services in connectionless mode are also planned.

Fibre optics seem to be the indispensable medium as far as transmission functions are concerned, capable of providing the capacity and performance necessary for high-speed services. Synchronous digital transmission, mentioned on p. 35, is the natural partner for the installation of broadband ISDN: the SDH system will be used as the network's physical layer, whereas ATM will optimize the use of these supports and be the basis for the services offered.

The worldwide installation of such a network can only happen progressively. Therefore, the way in which it interoperates with existing networks will be a prime factor in the success of broadband ISDN.

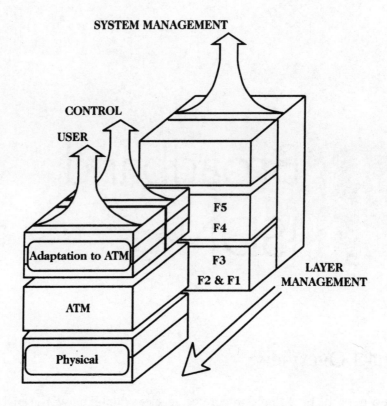

Figure 4.1 Separation into planes of functions

▧ 4.2 The architecture of broadband ISDN

The architecture of broadband ISDN is based on the concept of separate planes, which ensure the segregation of three groups of functions: user, control and management (Figure 4.1).

As well as being segregated into planes according to the stacks of protocols used, broadband ISDN is also based on a segmentation of the network that is largely based on its predecessor, narrowband ISDN (Figure 4.2), in which two subnetworks support the services:

- the distribution network between the subscriber and the local access switching unit, which may use a continuous flow of cells or synchronous digital frames;

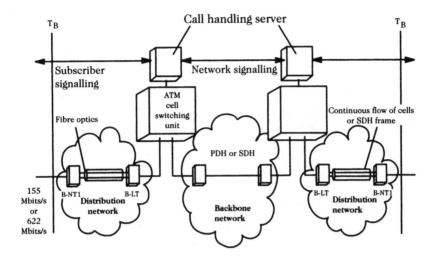

Figure 4.2 Broadband ISDN architecture

- the backbone network, which connects the local access switching units using synchronous digital transmission.

The segregation of the protocol stacks into planes and the segmentation of the network into subnetworks gives two reference models, which we will describe below.

4.2.1 The configuration reference model

A reference model is extremely useful when it comes to describing a complex environment. This model, which was inherited from narrowband ISDN, is made up of functional groups with points of reference between them, which may or may not be realized as physical interfaces.

Three main functional groups have been defined at the frontier between the distribution network and the domain of the user who is a subscriber to broadband ISDN services. Two of them are located one on each side of the access link:

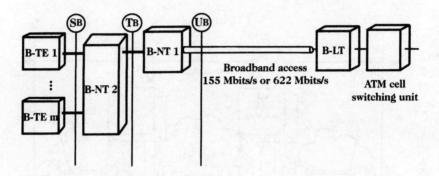

Figure 4.3 Connection to broadband ISDN

- switching unit side, the broadband line termination (B-LT), which connects the broadband switching unit to the optical link;
- subscriber side, the broadband network termination 1 (B-NT1), which connects the **subscriber installation** and is the last element in the distribution network.

The B-NT1 line terminating equipment connects the installation to the network using an interface at reference point T_B. The subscriber installation is made up of a group of terminals (B-TE, broadband terminal equipment) connected either directly to the network or via switching units (B-NT2, broadband network termination 2 or digital subscriber termination). Figure 4.3 shows the reference configuration for a broadband connection.

The digital subscriber termination (B-NT2) is the framework of the subscriber installation and covers a large number of functions:

- managing the interfaces at reference points S_B and T_B;
- managing shared supports, such as the local area networks;
- managing the signalling between the subscriber and the access switching unit;
- determining the traffic's profile and allocating resources;
- switching and multiplexing ATM cells;
- adapting the information flows to the ATM cell format;

- supporting internal communications and filtering the incoming
  and outgoing communications of the subscriber installation.

A broadband PBX (*private branch exchange*), a data flow multiplexer and
a communications controller are examples of the sort of equipment that
can accommodate this kind of functional group (B-NT2).

A broadband terminal adapter (B-TA) is used to connect terminals,
which are not compatible with the interface at reference point S_B.

4.2.2 The protocols reference model

We will better understand the division of functions in the environment
based on ATM if we look at the protocols reference model. Figure 4.1 on
p. 72 shows the relationship between the planes and the layers. This
model has three planes:

- the user plane, responsible for the information flow generated,
  with error detection and recovery when necessary;
- the control plane, which manages the calls and the connections
  they entail: it includes support for the signalling system used
  for connecting the subscriber (p. 79);
- the management plane, which contributes to the management
  of the various protocol stacks (**layer management**) and to system administration (**system management**).

This architecture is based on three shared layers:

- the physical layer, which directly depends on the transmission
  system;
- the ATM layer, which uses the physical layer's services and
  provides cell switching, multiplexing and routing functions;
- the adaptation layer, which provides a service to the flows
  coming from the control plane or the user plane.

The adaptation layer resides in a terminal (B-TE) or in a digital subscriber
termination (B-NT2). It is also found in the network switching units for
interoperating services, where it provides the interface with the control
plane.

4.2.3 Reference points

Reference point T_B is the frontier, at least in Europe, between the operator domain and the user domain. Reference point U_B should play this role in the context of the current liberalization that is opening up all subscriber equipment to competition, but because it has not been defined point U_B will not be discussed in this chapter. Beyond point T_B, the subscriber installation usually requires a distribution structure to reach the terminals. The following two notions will be discussed separately:

- the interface at reference point T_B;
- the subscriber's distribution structure in the context of broadband ISDN.

The interface at reference point T_B

Broadband ISDN uses a new generation of user network connections, based in principle on the use of fibre optics in the distribution network. The characteristics of the interface at reference point T_B are as follows:

- the interface may be electrical (coaxial cable) or optical (monomode fibre);
- the real bit rate is 155.520 or 622.080 Mbit/s;
- the interface's structure is made up of a continuous flow of ATM cells or G.709 synchronous frames;
- the signals are coded in CMI (coded mark inversion) for an electrical interface, or in NRZ (for an optical interface.

This set of variants means there are several interfaces at reference point T_B, given the possible choices between interface structures (synchronous frames or continuous flow of cells) and transmission technologies (electrical or optical).

Distribution structure

There are two modes for connecting several terminals to a broadband digital access:

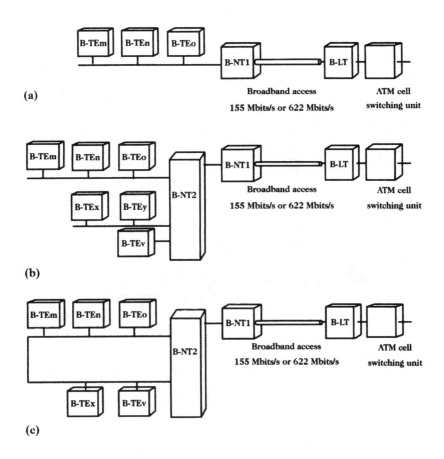

Figure 4.4 Subscriber distribution structure, (a) Bus; (b) star with B-NT2; (c) ring with B-NT2

- connecting each terminal (B-TE) to a single device that provides the switching functions;
- using a connection based on the distribution of the switching functions to the terminals, which are themselves distributed along a bus or ring.

These two connection modes give, in turn, three types of topologies for the subscriber distribution structure: star, bus or ring, as shown in Figure 4.4.

A shared transmission channel is used for the bus and ring configurations. The first 4 bits of the ATM cell's header are used at the UNI by the protocol that is responsible for equitable sharing of the transmission medium, thus allowing transfer of the traffic from each of the active terminals. These bits are called the GFC (generic flow control). There are two operational modes:

- **Uncontrolled mode**, in which the GFC bits are ignored on reception and set to zero on transmission, is used for **point-to-point configuration**;
- **Controlled mode**, in which two groups of protocols are under consideration: DQDB-type protocols and cyclic-type protocols. At present, no consensus has been reached on standardization and several proposals are still being studied. This mode is used for point-to- multipoint configuration.

These various structures coexist, because the choice between them is always the result of a technical/economic compromise depending, among other things, on:

- the number and type of terminals;
- the nature of the site (campus, tower block, isolated premises, for example);
- equipment management;
- the technological characteristics.

Whether the switching function is centralized or distributed, it always provides an interface at reference point S_B that is independent of the distribution structure (Figure 4.4). The definition of this interface is still the subject of standardization work. The debate is still open between identical characteristics to reference point T_B (the option chosen for narrowband ISDN) and much lower speeds, better adapted to the needs of the terminals.

■ 4.3 Signalling systems

Very flexible management of the bit rate assigned to each virtual connection is one of the main characteristics of the ATM technique. In particular, ATM allows a call set-up procedure that optimizes the use of the network's resources. This procedure can be divided into two phases:

- the establishment of a virtual connection between the terminal and the network, without allocation of the necessary resources;
- the allocation of resources, but only if the called terminal is both available and compatible with the calling terminal.

The allocation of resources is under the control of two signalling systems that cover the syntax and the semantics of the information exchanged between the user and the network or between nodes in the network. This information, which for a long time was limited to rudimentary signals, has now taken the form of structured messages on digital networks, especially on narrowband ISDN.

4.3.1 Signalling at the user network interface

The signalling system used between the user connection and the local exchange is an extended variant of the protocol used by narrowband ISDN, DSS-1 (digital signalling system-1). DSS-1 contains two levels of protocol: Q.921 and Q.931. The data link protocol, Q.921, which transports the Q.931 format signalling messages, is not good enough for high-speed connections and has been replaced by a protocol that is better adapted to that environment. This new protocol, called SSCOP (service-specific connection-oriented protocol or Q.2100), is based on AAL type 5 and forms with it an adaptation layer called SAAL (signalling ATM adaptation layer). Figure 4.5 shows the relationship between the signalling and adaptation layers.

SSCOP is used to transfer, in an assured mode, signalling messages, which are variable-length data units. In addition to a flow control mechanism, SSCOP includes means for detecting and selectively retransmitting missing data units. This assured service greatly simplifies the implementation of signalling protocols.

Thus, SSCOP completes the service provided by the underlying layer: because this layer does not implement error handling, the service provided by AAL type 5 corresponds to a non-assured transfer mode, which is only able to detect erroneous data units by means of the CRC 32 field.

High speeds make the classic 'go back on error' protocol (**go back-N**), used in HDLC-type protocols, inefficient: the error recovery procedure proposed by SSCOP is a variation of an HDLC mode called **balanced check mode**.

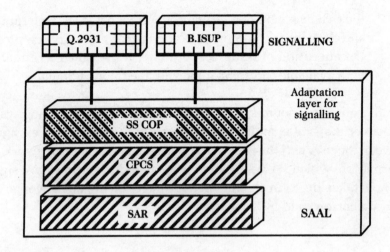

Figure 4.5 Relationship between the signalling systems and the adaptation layer

SSCOP is a connection-oriented protocol and provides a set of messages to establish and release a connection: BGN (**Begin**), BACK (**Begin Acknowledge**), BREJ (**Begin Reject**), END and ENDAK (**End Acknowledge**). These messages have a role similar to commands or responses such as SABM, UA, DISC and DM used in the HDLC protocol.

Once the SSCOP connection is set up, data units are transferred by means of SD (**Sequenced Data**), POLL, STAT (**Solicited Status**) and USTAT (**Unsolicited Receiver State**) messages.

The transmitter sends data units using SD-type messages, and provides the number of the next message to be sent through a POLL command. The receiver answers POLL by a STAT-type message indicating which was the last valid SD-type message received. It can also, on its own initiative, answer POLL through an USTAT-type message, the format of which is similar to the STAT message.

As an example, after transmitting four SD-type messages numbered 1, 2, 3 and 4, the transmitter can send a POLL(5) message to indicate the number of the next message. If there was no transmission error, the receiver sends back a STAT(5) message. If messages 1 and 2 only were correctly received, the receiver would send STAT(3)[3,5]. to request a retransmission of messages numbered 3 and 4.

The Q.2931 signalling protocol is derived from Q.931 and takes the characteristics of broadband services into account: a separate virtual connection is used for the signalling between the subscriber access and the local exchange.

Schematically, a Q.2931 connection call consists in sending a SETUP message, which includes the destination address, the required quality of service, the AAL type to be used and the source traffic characteristics. If the network can accept this new connection, it answers the source through a CALL PROCEEDING message, which provides the VPI and VCIs to be used for this new connection, and forwards the SETUP message towards the target destination. After receiving it, the receiver checks that it can accept the proposed parameters and then answers through a CONNECT message; otherwise it rejects the connection call using a RELEASE message. A RELEASE message is also normally used to close a connection when the communication is over.

4.3.2 Signalling at the network-to-network interface

With the advent of narrowband ISDN, a specific system for signalling between network nodes became generalized: signalling system no. 7 (SS-7). This internal network signalling system has also been modified to take into account the characteristics of broadband services. The ISUP protocol (integrated service user part or Q.761/4), equivalent to Q.931 for SS-7, has been extended to become B-ISUP (Q.2761/4).

The SSCOP, which is already used to protect the integrity of the data transferred between the subscriber and the network, using the Q.2931 protocol, is also used with the B-ISUP protocol. Re-employing the techniques implemented for narrowband ISDN has two significant advantages:

- it capitalizes on the experience gained with narrowband ISDN in the context of signalling;
- interoperation between narrowband and broadband ISDN is easier.

4.3.3 Metasignalling at the user network interface

The implementation of the virtual connection between the subscriber and his/her access exchange depends on the configuration of the user installation. In a **point-to-point configuration**, where a single device

(B-NT2 or B-TE) is connected to the interface at reference point T_B, a permanent virtual connection is used: virtual channel number 5 on virtual path number 0.

On the other hand, if the configuration allows several devices to share the digital access (**point-to-multipoint configuration**), a **metasignalling** procedure is required to manage the configuration. The exchange occurs on a metachannel which is virtual channel number 1 (on virtual path number 0). Furthermore, a broadcast signalling virtual channel (BSVC) is used to present an incoming call to all the devices in the 'multipoint' configuration so that they can determine their level of compatibility with the calling device. This broadcast channel is identified by virtual channel number 2 on virtual path number 0. Therefore, there is one metachannel and one broadcast channel per interface. This number can be increased by changing the virtual path number.

The metasignalling procedure (Q.142X) is similar to that of assigning an identifier to a terminal (TEI, terminal end-point identifier) on the narrowband ISDN passive bus. As in the case of narrowband ISDN, it is managed by the management plane and not by the control plane. Its main functions are the following:

- setting up, releasing and checking the status of virtual signalling channels;
- resolving contentions in the allocation of virtual signalling channel and path identifiers;
- managing the bit rate assigned to the virtual signalling channels.

■ 4.4 Services offered by broadband ISDN

Once again, we go back to the principles of ISDN: bearer services, teleservices and supplementary services. The bearer services are based on the asynchronous transfer mode, which handles any type of information flow as a continuous succession of ATM cells, and are complemented by the appropriate adaptation functions. This architecture means that the following bearer services can be offered on the broadband interface:

- a virtual circuit service, permanent or on request, in which the bandwidth is reserved (circuit emulation);

- a virtual circuit service, permanent or on request, in which the bandwidth is allocated statistically (equivalent to packet switching);
- a datagram service based on E.164 addressing.

In general, these services already exist, but only on specific networks with limited speeds. When they become available on a single interface it should make multimedia communication easier. However, given the amount of investment necessary, the introduction of broadband ISDN will be gradual. Any operator introducing a service will have to reconcile two contradictory objectives:

- a service can only be developed if it is based on a network with a large geographic coverage;
- initial investment cannot exceed the real demand if it is not to compromise the development's financial viability.

Thus, the broadband ISDN start-up phase can only be based on services for which the demand is both limited, in terms of the population, and restricted, from a geographical point of view.

▓ 4.5 Deployment of the ATM technology

The telecommunications operators are deploying a new network infrastructure based on ATM for the switching function and on SONET (in the United States) and SDH (in Europe) for the transmission.

The first services that are offered consist in CBR or VBR permanent virtual circuits, available on interfaces that depend on the operators. In many countries, a connectionless SMDS/CBDS (switched multimegabit data service/connectionless broadband data service) based on a DQDB platform was proposed before ATM services (e.g. in the UK). Such a service can also be proposed on an ATM platform using a specialized server (e.g. France Telecom). Thus, the SMDS/CBDS can be provided under three forms (Figure 4.6).

- Direct connection to an SNI (subscriber network interface) in the case of a DQDB platform, or to a UNI for an ATM platform. In this case, the service corresponds to a connectionless data transport service.

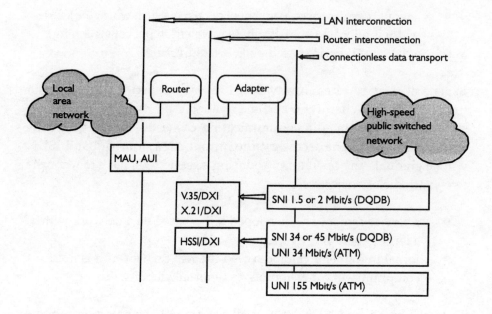

Figure 4.6 SMDS/CBDS service

- Connection of a router through an adapter that converts the SNI/DQDB or the ATM/UNI interface into a DXI (*data exchange interface*). The latter uses a V.35 or X.21 physical interface at 1.5 or 2 Mbit/s and HSSI (high speed serial interface) at 34 or 45 Mbit/s. Such a connection corresponds to a router interconnection service.
- Connection to an AUI (attachment unit interface) or MAU (medium attachment unit) for a local area network interconnection service.

4.5.1 ATM in Europe

Many European public operators signed an MoU (memorandum of understanding) to interconnect their national pilot networks. An ATM digital cross-connects network has been built since 1994. High-speed services have been evaluated on this platform, such as circuit emulation, frame relay service and connectionless data service.

The European pilot network has also been used to validate ETSI standards (e.g. ATM cell mapping mode onto a 34.368-Mbit/s PDH payload) and the Eurescom specifications. This research institute, created by 26 public operators, and based in Heidelberg, helps in deploying three technologies in Europe: ATM, TMN (telecommunication management network) and IN (intelligent network).

The pilot network has made possible the interconnection of research centres in the countries that signed the MoU, as well as pilot users connected to a broadband national platform. The offerings proposed by BT, Deutsche Bundespost Telekom and France Telecom characterize the high-speed switched services available in Europe, and result from the experience these operators have developed on the pilot network.

BT, Deutsche Bundespost Telekom and France Telecom propose an SMDS/CBDS at 2, 4, 10, 16 and 25 Mbit/s. Depending on the operator, the tariffs comprise the following items:

- an inclusive subscription, according to the rate, and which does not depend on the distance (BT and France Telecom);
- a charge that depends on the volume of information carried (Deutsche Bundespost Telekom and France Telecom);
- a router placed on the customer's premises whose cost depends on the type and the number of ports used to deliver the service.

As an example, the inclusive tariff proposed by BT is an interesting competitor of high-speed leased links. Similarly, France Telecom's CBDS (Transrel ATM) volume charge is much lower than Transpac's.

Deutsche Bundespost Telekom proposes CBR ATM virtual circuits, in addition to the SMDS/CBDS. The tariff comprises an inclusive subscription fare that depends on the access rate (2, 34 or 155 Mbit/s) and a charge according to the call duration and the distance. Three distance zones have been defined: local, up to 250 km and beyond 250 km. The service is charged for each hour of connection.

4.5.2 ATM in the United States

The American operators fall into two main categories: long-distance operators (IXCs, interexchange carriers) and local operators (LECs, local exchange carriers). AT&T, MCI and Sprint are part of the first group, together with newcomers such as MFS Datanet and WilTel (now called

LDDS WorldCom), which all provide commercial ATM services. In the second group, the RBOCs (regional Bell operating companies), such as Ameritech, Bell Atlantic, BellSouth, Nynex, Pacific Bell and US West, also offer ATM commercial services. These services depend on the type of access, the type of service and the tariff.

Practically all the carriers use DS3 (45 Mbit/s) and OC3c (155 Mbit/s) interfaces to connect users to the ATM services, and many of them also propose a DS1 (1.5 Mbit/s) interface. All operators provide CBR and a VBR service on point-to-point virtual permanent connections. The difference resides in the offering granularity:

- Bell Atlantic proposes a 0.5 Mbit/s service granularity on a DS3 interface and a 1 Mbit/s service granularity on OC3c;
- BellSouth has another granular offering (1.5, 4, 10, 16, 20, 45, 100 and 155 Mbit/s).

In addition to the ATM services, some operators provide equipment placed on the customer's premises to emulate conventional services over an ATM network (circuit emulation, frame relay service, LAN interconnection service...). The interfaces proposed to the users are then the following:

- DS1 or DS3 for circuit emulation;
- V.35/Q.922 for frame relay;
- AUI, MAU or MIC (media interface connector) to interconnect LANs;
- DXI to interconnect routers.

The case-by-case tariffs for such services depend on the type of interface, the number of sites interconnected and on additional features (e.g. network management function). Most of these operators are part of the first group (IXC). The catalogue-based tariffs from the LECs usually include a monthly fee that is dependent on the access rate and a variable charge as a function of the virtual connection rate and the type of service. This principle applies for both CBR and VBR services, except for AT&T, which includes a distance-based charge for CBR traffic, and BellSouth, which takes the distance into account in the access fare (distance between the user and the ATM switch). The rate for CBR traffic is usually higher than for VBR. Similarly, the rate for an OC3c interface is typically three times

Table 4.1 ATM services in the United States

Operator	Tariffs	Principles	ATM interfaces	Other interfaces
AT&T	Catalogue	Fixed (subscription) Usage (VBR) Distance (CBR)	DS3, DS1 (planned)	
GTE	Case by case		DS3, OC3c	
MCI	Case by case		DS3, OC3c, DS1 (planned)	
MFS Datanet	Case by case		DS3, OC3c	V.35/FR, AUI, MAU, MIC, DXI
Sprint	Catalogue	Fixed (subscription) Use (VBR, CBR)	DS3, DS1 (planned)	
WilTel	Case by case			Channel, AUI, MAU, MIC, DXI
Ameritech	Case by case		DS1, DS3, OC3c	
Bell Atlantic	Catalogue	Fixed (subscription) Use (VBR,CBR)	DS3, OC3c	
BellSouth	Catalogue	Distance (subscription) Use (VBR, CBR)	DS3, OC3c	
Nynex	Case by case		DS3, OC3c	
Pacific Bell	Catalogue	Fixed (subscription) Use (VBR, CBR)	DS3, OC3c	
US West	Catalogue	Fixed (subscription) Use (VBR, CBR)	DS3, OC3c	

Table 4.2 ATM operators in the United States

Operator	Internet
AT&T	www.att.com
GTE	www.gte.com
MCI	www.mci.com
MFS Datanet	www.mfsdatanet.com
Sprint	www.sprint.com
WilTel	www.wiltel.com
Ameritech	www.ameritech.com
Bell Atlantic	www.ba.com
BellSouth	www.bellsouth.com
Nynex	www.nynex.com
Pacific Bell	www.pacbell.com
US West	www.uswest.com

as much as the one for a DS3 interface. Table 4.1 summarizes the main characteristics of the services proposed by the two groups of American carriers, and Table 4.2 indicates how to contact those operators.

5 | ATM and private networks

■ 5.1 Introduction

The use of ATM technology in public networks has been described in the context of the broadband ISDN (p. 71). Two types of interface are used in that environment: the UNI (I.432), which connects users to the services offered by the public networks, and the NNI, which interconnects public networks. The Q.2931 signalling system used at the UNI is different from B-ISUP used at the NNI (p. 81). These two signalling protocols use the same addressing structure to identify the source and the destination of any connection: the E.164 addresses that are widely used in public networks, in particular for the telephone service, which concerns almost a thousand million subscribers.

This set, including I.432, Q.2931, B-ISUP, E.164 and other related standards, has been defined by ITU-T to fulfil the public networks' needs. Private networks, either local or wide area, have some characteristics in common with public networks, but they also have divergent aspects. This is why the ATM Forum was created. This body has redefined, for a private network environment, the UNI and NNI interfaces, the signalling systems and the related addressing structure. In particular, as indicated in Figure 5.1, the P-NNI (private NNI) protocol applies between two switches as well as between two private networks.

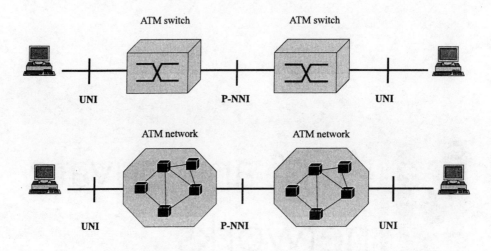

Figure 5.1 UNI and NNI interfaces

■ 5.2 UNI signalling

The ATM Forum defined the UNI 3.1, which is a slight evolution of version 3.0, and works on a new version called 4.0. The signalling system is a simplified version of Q.2931, as the private networks do not have to take into consideration all interworking cases met in the context of public networks. The UNI 3.1 signalling system uses fewer messages and thus has fewer operational states. In addition to the limited number of messages for point-to-point connections, this system includes specific messages to handle multipoint connections. As for the Q.2931 protocol, the UNI 3.1 signalling system is carried on virtual channel number 5 (VCI = 5) on virtual path number 0 (VPI = 0).

The method to set up a connection is the one used in all public telecommunication networks: the connection call is propagated from the source towards the destination and a series of links are concatenated to build a connection. The way in which the call is routed depends on the destination address, the quality of service required by the source and on the traffic characteristics. The destination station can either accept or reject the proposed connection, depending on negotiable criteria.

▓ 5.3 Addressing structures

Beyond the simplification brought to Q.2931 signalling system by the ATM Forum for point-to-point connections, the main contribution deals with the extended addressing structures handled by the UNI 3.1. In fact, the E.164 addressing range is a resource of the public domain that cannot easily be extended to the private domain. Thus, the ATM Forum defined a structure based on the syntactic principles defined by ISO for the NSAP (network service access point). The addressing range has a fixed 20-byte length and is divided into three parts: the AFI (authority and format identifier), which defines the type and the format of the IDI (initial domain identifier), which in turn indicates which body is responsible for address allocation; and the DSP (domain-specific party), which contains the address needed for routing. There are three possible formats:

- the E.164 format encoded into an NSAP: the IDI field is then the E.164 address;
- the DCC format, in which the IDI field contains the DCC (data country code) to identify the countries in accordance to ISO standard 3166: the addresses are then handled by the ISO national authorities;
- the ICD format, for which the IDI field is an ICD (international code designator) handled by BSI (British Standards Institution) in accordance with ISO standard 6523.

Figure 5.2 illustrates these three formats used in the domain of private networks.

Compared with ISO NSAP, the addressing structure proposed by the ATM Forum combines the **routing domain** and **area** identifiers into a single field called HO-DSP (high-order DSP). The ESI (end system identifier) field identifies the stations: this field usually contains the station MAC address, as defined by the IEEE for local area networks.

The ATM Forum defined a self-configuration procedure that allows a station and an ATM switch to exchange, through the UNI 3.1 interface, the address elements they have at connection time. This procedure uses the ILMI (interim local management interface) protocol. The station registers with the ATM switch and sends to it the contents of its ESI field; the switch returns its own ATM prefix (or **zone**). At the end of this dialogue, the switch has recorded the presence of a new station characterized by

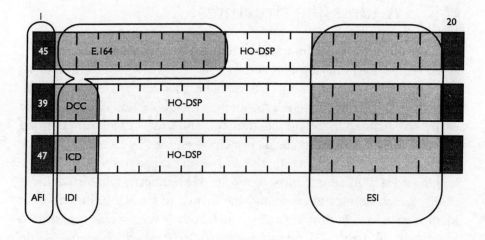

Figure 5.2　ATM addresses in private networks

the ESI field contents, and the station now has a complete ATM address, comprising its own identification and the ATM network's prefix, received from the switch. Figure 5.3 illustrates the registration procedure.

■ 5.4 Protocols at the NNI

An addressing structure specific to ATM implies that any network based on that technology appears as a subnet for the protocols that have their own addressing structure. This situation is not really new and already exists, e.g. between IP (Internet protocol) and X.25. When an X.25 network is used within an IP network, an association is needed between the IP addresses and the X.121 addresses, as X.121 defines the addressing structure for X.25 networks. This association, or address resolution, is realized at the boundary of the two networks. Such a principle dissociates the development of ATM from that of the other network protocols and allows them to evolve independently.

Inside an ATM network, a routing protocol through the NNI is needed to carry the signalling messages dealing with the setting up of the connections and, once those connections are established, to support the data

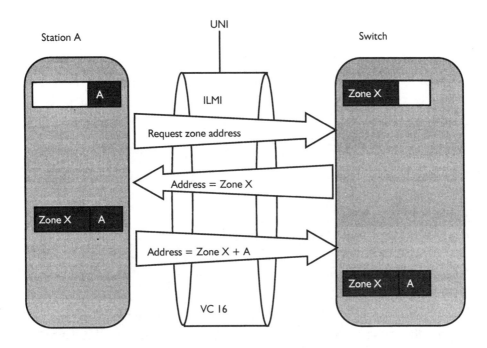

Figure 5.3 ILMI registration protocol

messages. This protocol plays a role equivalent to the RIP (routing information protocol) or OSPF (open shortest path first) protocol used in the IP-type networks. The ATM Forum is working on the definition of the P-NNI interface, which includes such a routing protocol. As far as the public networks are concerned, they will use another interface identified by the ATM Forum as the B-ICI (broadband intercarrier interface); this interface capitalizes on the ITU-T standards, mainly on the B-ISUP (p. 81) protocol, which in turn uses the MTP (message transfer part) protocol for the routing functions.

The P-NNI interface includes two protocols: a signalling protocol and a routing protocol. The former is a variation of the one used on the UNI 3.1 interface and adds information elements specific to the P-NNI environment. The routing protocol allows the determination of the route to be used by the signalling messages and, later, by the information transfer messages, when the connections are set up. P-NNI phase 0 is limited to a

Table 5.1 Signalling and routing protocols

	Addressing	UNI signalling	NNI signalling	Static routing	Dynamic routing
Private network (ATM Forum)	NSAP-ATM	UNI 3.1	IISP P-NNI phase 1	IISP	P-NNI phase 1
Public network (ITU-T)	E.164	Q.2931	B-ISUP	MTP	

static routing, whereas P-NNI phase 1 allows a dynamic routing function. To avoid any confusion, P-NNI phase 0 has been called IISP (interim interswitch signalling protocol).

Table 5.1 illustrates the different signalling and routing protocols used by the private and the public networks.

5.4.1 Principle and goal of the routing

The routing tables can be built either manually or automatically. In manual mode, the network administrator generates the tables by means of a network management tool. In automatic mode, the routers, or the routing function integrated within the switches, discover the topology through routing protocols. These protocols use two types of algorithms based on the number of intermediate links (**distance vector**) or on various criteria such as the cost of a link as a function of its data rate (**link-state**). The IP networks use RIP or OSPF. RIP is a 'distance vector'-type protocol, for which the distance is expressed by the number of intermediate nodes, whereas OSPF is a 'link-state' protocol.

Building the routing tables results from a self-learning process that leads to a complete description of the network, after a convergence time. The 'link-state' protocols converge more rapidly than the 'distance vector' protocols. Besides the type of algorithm used, the routing protocols differ by their scope of action: they are either intradomain or interdomain protocols. RIP and OSPF are intradomain protocols, whereas others, such as BGP (border gateway protocol) or EGP (exterior gateway protocol), are interdomain protocols. These protocols are used in networks built up of routers, each of them being in charge of two functions:

- it determines how to reach the destination stations and the nodes of the network it belongs to by periodically exchanging information with its neighbour routers;
- it switches the received packets towards the destination indicated in the packet's header according to the acquired routing information.

In a network built up of switches which use a dynamic routing algorithm, the ATM switch must also have information on the state of the routes in the network to which it is connected. The difference with a router concerns the goal itself:

- the router operates in a connectionless environment in which the routing information is directly used to switch each packet individually;
- the ATM switch works in a connection-oriented mode and the routing information allows a connection to be set up that will then be used between a source and a destination.

5.4.2 Routing protocol P-NNI phase 1

One of the major characteristics of ATM is the ability to guarantee a quality of service for the duration of the connection that is set up through the network topology. Four qualities of service (p. 23) are proposed: constant bit rate (CBR), variable bit rate (VBR), available bit rate (ABR) and unspecified bit rate (UBR). The public networks presently limit their offering to CBR and VBR. The network guarantees the requested quality of service according to a set of criteria that is different for each service (e.g. cell loss ratio, cell transfer delay; p. 25). This guarantee is assured by the access switch through an access control mechanism (p. 25). This mechanism is fed by information collected in the whole network and carried by specific messages called PTSPs (P-NNI topology state packets). It makes great use of the OSPF routing protocol but, to take into account the large number of services proposed by ATM technology, the set of parameters used to determine a route is much larger.

Thus the P-NNI phase 1 routing protocol belongs to the 'link-state' family. However, it is functionally richer as information elements about the status of the switches have been added to the link-state parameters. These parameters take into account the maximum cell transfer delay, the cell delay variation and the maximum cell loss ratio of priority cells (CLP = 0) for CBR and VBR traffic. These parameters' attributes mainly define the available rate, expressed as the number of cells per second.

From this set of parameters, the source switch builds up a DTL (designated transit list) in several steps:

- the links that cannot afford the required rate are eliminated;
- a route calculation is made on the limited list resulting from the first step;
- as the previous step may lead to several possible routes, the cumulative attributes, such as the delays, are evaluated to determine the best route (at this moment) to reach the destination;
- once the route has been determined, it is translated into a DTL and placed in the connection set-up message.

Each intermediate switch operates an admission control that may lead to a rejection status, as the traffic conditions may have changed between the time the source switch determined the route to be used and the time when the request is actually presented. To avoid too many call set-up rejections, the route is calculated again by the switch located upstream with respect to the experienced blocking state. This restart mechanism is called '**crankback**'.

■ 5.5 ATM in local area networks

5.5.1 Limitations of current LANs

The techniques and protocols that are most often used in the current local area networks (contention buses, token rings, FDDI networks, bridges, routers) are widely deployed, as there are more than 50 million Ethernet or token ring adapters in use today. This number is continuing to increase owing to the moderate cost for that type of connection, the large number of available suppliers and applications and the interworking capability brought by mature standards.

However, the new applications suffer from two limitations inherent to conventional LANs: the rate offered to every station is increasingly proving to be insufficient and support of isochronous traffic is almost impossible.

- The diversity of information to be handled has changed radically: whereas information was purely alphanumeric at the time of the first LANs, it is now necessary to deal with graphics and images, either fixed or animated. In addition, the processing capability justifies more convivial applications and the data processing structures and the network architectures are also changing: data processing systems based on large computers have given way to a more distributed processing approach that interconnects all an enterprise's LANs. Finally, the client server architectures generate large flows of information. For all those reasons, the rates are increasing and have reached the limits of the current LANs.

- Conventional LANs were not designed to support isochronous traffic, which is now necessary to integrate voice, data and video into multimedia applications, and which induce additional constraints in synchronization and jitter.

The current LANs (also called '**legacy LANs**') cannot easily bring a large rate increase and accept isochronous constraints because of the shared-media principle, which was in fact one reason for their success.

The conventional MAC sublayer protocols propose a connectionless service for stations accessing a shared support, without the need for a complicated switching or connection set-up mechanism. They regulate access to the shared support whether or not through deterministic techniques, and this leads to limitations in performances and access times when the number of stations is increased.

In addition, support of isochronous traffic requires low but also strictly stable response times: such constraints are not compatible with the principle of carrying variable-length frames, as the transmission of a large data file may, for example, delay video frames beyond acceptable limits.

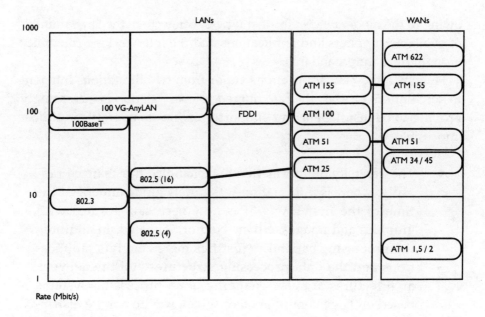

Figure 5.4 Evolution trends in local area networks

At the cost of an additional complexity (handling several priorities on a token ring, determining the token rotation time on a FDDI network), the conventional LANs can, at best, guarantee an upper-bound value for the response time.

As sharing of transport capacity increasingly becomes difficult, a possible approach consists in dedicating a LAN to a small number of stations with the same characteristics. Such a segmentation, which implies a more complex network administration, leads to an increase in the number of LANs and the interworking units (bridges or routers), which may in turn significantly reduce the overall performance.

5.5.2 Possible improvements

Currently, several technical solutions are being proposed. As far as possible, they preserve the existing investments, in terms of cabling, workstations and applications (see Figure 5.4).

MAC frame switching

As wiring concentrators (hubs) are widely used in both token ring and 10BaseT Ethernet LANs, a MAC frame-switching function is possible in these hubs: instead of providing a multiport repeater function (Ethernet) or a ring continuity function (token ring or FDDI), the hub directly switches every MAC frame towards the destination station. A self-learning mechanism is needed to discover the MAC address of the stations connected to each port. Filtering functions are also feasible with such equipment. Such an approach, which involves no change in the cabling or workstations, and which can be seen as an alternative solution to multiport bridges, allows several connections to be set up simultaneously.

There are two types of MAC frame switches: some wait until they have received a MAC frame completely and checked its validity before shipping it towards the appropriate port, whereas others switch the frames immediately ('on the fly') as soon as they have received and recognized the destination address field. The latter leads to very short switching times (about 50 μs), but with the risk of carrying invalid frames.

LAN segments comprising several stations can be connected to such switches; it is also possible to have a port dedicated to a single station (a server, for example). In the case of Ethernet buses, the probability of collision is thus reduced or nil. On a dedicated connection, the protocol can operate in duplex mode, at the cost of installing in the station an adapter that supports such a mode of operation.

Increased access rate

Another approach, which can also be combined with the MAC frame-switching technique, consists in increasing the access rate and designing hubs that support several links working at different rates. Such solutions keep applications unchanged but require the installation of a new adapter in the station. Several combinations allow existing cabling to be reused either on two category 5 UTP pairs ('data' quality) or on four category 3 STPs ('voice' quality). Remember that category 3, 4 and 5 cables are specified for a bandwidth of 16, 20 and 100 MHz respectively. Two techniques used at 100 Mbit/s, 100BaseT and 100VG-AnyLAN, directly compete with the copper version of FDDI on twisted pairs, also called CDDI (copper distributed data interface).

- The **100BaseT** technology (IEEE 802.3) maintains, at 100 Mbit/s, the principles used in 10BaseT Ethernet LANs: the protocol used to detect the collisions, the format of the frames and baseband transmission. As the data rate is 10 times larger, the stations furthest from each other on a segment can be separated by about 250 m (10 times less than on a 10BaseT network). The connection is normally done through two category 5 UTPs (100BaseTX), but a connection using four category 3 STPs (100BaseT4) is also possible. Furthermore, a connection by a multimode optical fibre (100BaseFX) has also been defined.
- The **100VG-AnyLAN** technology (IEEE 802.12) uses a new protocol, which is compatible with the ones used on Ethernet and token rings. This protocol, called DPAM (demand priority access method), is deterministic and can cope with two levels of priority, allowing flow from some critical applications to be favoured. As for the 100BaseT technology, the connection is realized through two category 5 UTP pairs or four category 3 STPs.

These various techniques described above, such as switched Ethernet or token ring, 100BaseT and 100VG-AnyLAN, are currently used to solve the problems experienced by some users who need high-performance LANs, but they have limitations that preclude them from being used within a high-speed network able to integrate voice, data and video flows. Moreover, ATM is considered as the leading future technology for both local and wide area networks: in the context of the LANs, ATM brings the advantages of the new solutions described above, in addition to other specific improvements.

5.5.3 LAN/ATM interworking

The ability to interoperate between LANs and ATM equipment is mandatory owing to the large number of local area networks installed. The interworking principle consists in using ATM as a subnet technology for existing protocols, such as IP. There are two addressing structures outside ATM: the network layer addresses (IP addresses, for example), and the IEEE MAC addresses. This duality leads to two interworking modes, as the ATM subnet has its own addressing structure (p. 91):

- an association between MAC addresses and ATM addresses, which thus operates the ATM technology as a MAC sublayer;
- an association between network layer addresses (such as IP addresses) and ATM addresses, which then uses ATM as a logical link with respect to the network protocols.

These modes of association have led to two standards: **LAN emulation** defined by the ATM Forum, and **IP over ATM**, defined by the IETF (Internet Engineering Task Force). In both cases, the virtual connections needed between the source and the destination are set up using the UNI 3.1 protocol defined by the ATM Forum (p. 90).

LAN emulation

The addressing formats are not the only difference between ATM and the local area networks. ATM is a technology based on the fact that a connection exists between the source and the destination, whereas LANs operate in a connectionless mode. To behave as a LAN, as seen from the upper-layer protocols, and to provide them with an apparent connectionless service, the interworking layer has to emulate a LAN, and this explains the name LAN emulation (LANE). This emulation function presents to the network layer protocols a service interface that is the same as the one given by the LAN protocols. On the other hand, as far as the physical layer is concerned, the shared medium, which is the LAN basic function, is replaced by an ATM (sub)network. The frame broadcast, which is a native LAN function, must be solved through specific means in the context of a switched network.

The LAN emulation protocol is thus placed between a network layer protocol and the AAL. It is transparent for the ATM layer and is located in the terminal stations at the ATM network boundary. It can be implemented on an interface card that directly operates the most common LAN interface drivers, such as NDIS (network driver interface specification) or ODI (open data-link interface), or it can be integrated into an ATM bridge that guarantees LAN/ATM interworking function (see Figure 5.5).

The LANE protocol must be able to work in a segmented LAN environment and thus learn where the MAC addresses are located on the segments. This protocol uses a client server mode and emulates a token ring or a contention bus. It comprises two elements: the LEC (LAN emulation

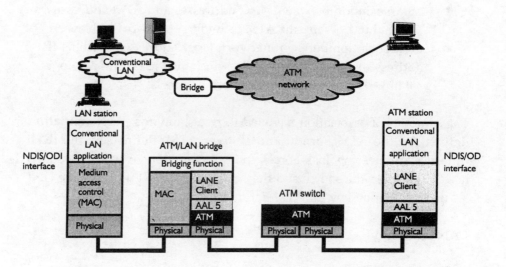

Figure 5.5 LANE protocol layers

client) and the emulation service which, in turn, may consist of several servers:

- the LES (LAN emulation server), which registers the clients of an emulated LAN and associates the MAC addresses and the ATM addresses;
- the BUS (broadcast and unknown server), which handles the sending of frames to all stations (**broadcast**) or to some of them (**multicast**) and propagates the first frames from the stations when the address association is not yet completed;
- the LECS (LAN emulation configuration server), which is responsible for the configuration of the LANs within a given domain.

An emulated LAN is thus formed by associating clients (LEC), identified by their MAC and ATM addresses, and two specialized servers (LES and BUS). The table that associates LEC, LES and BUS resides in the LECS. Figure 5.6 illustrates an emulated LAN (ELAN). Two types of clients can be seen: a station with a LANE function card, and an ATM/LAN bridge (also called a 'proxy' bridge) in charge of interworking with the conventional LAN stations.

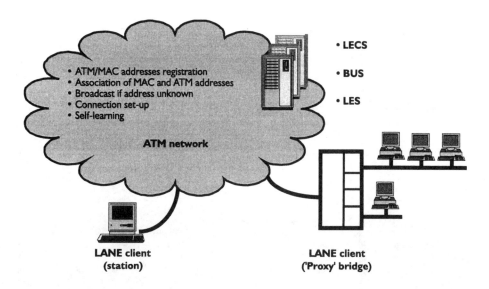

Figure 5.6 LANE clients and servers

For an emulated LAN, the protocol specifies the relation between the clients and the three LANE servers (LES, BUS, LECS). It is thus a UNI-type protocol called LUNI or LAN emulation user network interface (see Figure 5.7). Presently, the relation between LANE services (LNNI, LAN emulation node-to node-interface) is not totally standardized.

When it is initialized, a client needs to get its full ATM address, through the registration procedure described on p. 91. The client then establishes a virtual connection with the LECS. Several mechanisms allow the client to know the LECS ATM address:

- by consulting the ATM switch, through the ILMI protocol;
- by using a reserved ATM address;
- by means of a dedicated permanent connection (VPI = 0, VCI = 17).

The LECS returns to the client the ATM address of the LES which is in charge of the emulated LAN the client has to join, together with the type of this LAN and various parameters, such as the maximum packet size.

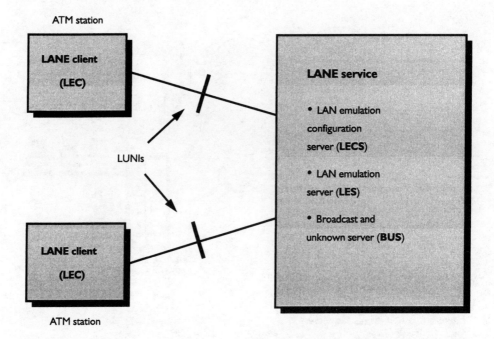

Figure 5.7 LUNI interface

Once the LES ATM address is known, the client registers with it by giving its ATM and MAC addresses: if it is the case, the client also identifies itself as 'proxy'. The LES, in turn, provides it with an identification number.

Upon request from a source station, the LES gives the ATM address of a destination station for which the source knows only the MAC address. As all clients have registered with it, the LES should be able to give such an address at any time. The same procedure is used by the clients to obtain the BUS ATM address.

In some situations, however, the LES does not know the address association when it receives the request (called LE-ARP request or LAN emulation address resolution protocol request), particularly when the destination is a conventional station connected to an ATM bridge ('proxy' client). The LES then propagates that request towards all 'proxy' clients of the emulated LAN. The time needed to determine the destination ATM address may exceed the time-out values set by the source client: to avoid any loss of data if the time-out is exhausted before the association is

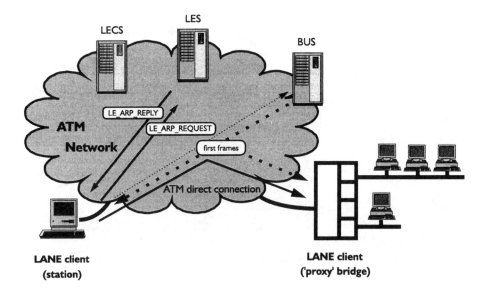

Figure 5.8 LANE address resolution

resolved by the LES, the client sends its first frames to the BUS. The latter server broadcasts these frames towards all the stations of the emulated LAN logically connected to it. This procedure is similar to the conventional LAN bridging techniques (Figure 5.8).

This dual procedure comprising the LES resolving the destination ATM/MAC address association and the BUS broadcasting the first frames may lead to a situation in which the address association is given to the source client while it is already using the BUS to send frames. As it now has a direct virtual connection to the destination client, the source client stops sending to the BUS and releases the connection used via the BUS. This purge procedure prevents the destination client receiving abnormal frames (duplications or sequence errors).

The LANE service makes use of a set of direct ATM connections between the clients and the servers (Figure 5.9). The first three are control connections, whereas the others are used to carry data:

- a bidirectional configuration connection between a LEC and its LECS (**configuration direct VCC**) that can be released once it has been used;

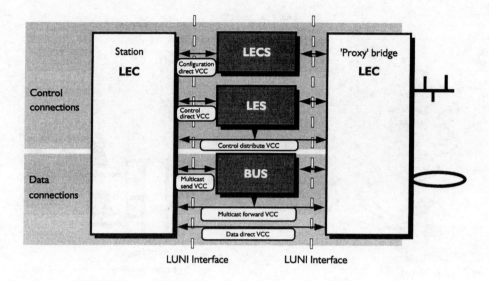

Figure 5.9 Virtual connections for LANE service

- a bidirectional control connection (**control direct VCC**) between a client and its LES;
- an optional unidirectional connection (**control distributed VCC**) between the LES and its clients to broadcast control messages;
- a bidirectional connection (**multicast send VCC**) between a LEC and its BUS, so that the client's frames can be broadcast by the BUS;
- a unidirectional connection (**multicast forward VCC**) for the BUS to broadcast data to its clients;
- bidirectional connections (**data direct VCC**) to exchange data with the other LANE clients.

Virtual LANs

An emulated LAN (ELAN) behaves as a conventional LAN without capacity limitations. It also allows virtual LANs to be built. Thanks to the LECS, the network administrator can define several emulated LANs on the same ATM network. The clients can be assigned or reassigned to any

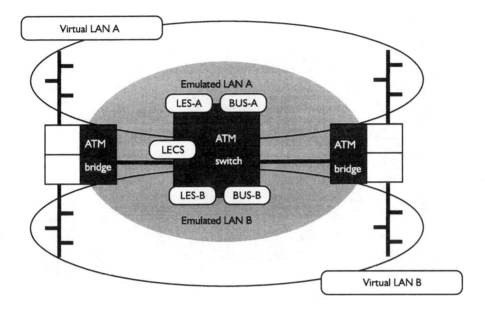

Figure 5.10 Virtual LANs

virtual LAN independently of their physical location. This split between the physical topology and the belonging of a station to a logical network allows the local area network to be viewed as a virtual space (see Figure 5.10). The move from a virtual LAN to another is simply made by transmitting the ATM and MAC addresses from one ELAN to the other. The LECS holds information about which station belongs to which ELAN. This concept of virtual LAN is not restricted to the stations (LECs) connected to an emulated network in ATM mode, it can also be extended to conventional LAN segments connected to the emulated network by a multiport bridge.

So the principle of the virtual LANs is similar to the conventional bridging techniques used to interconnect physical segments, and thus it does not allow direct communication between virtual LANs, unless these virtual LANs are connected by a routing function (Figure 5.11). Such a function must include a dedicated port for each interconnected ELAN or a single ATM interface supporting a virtual connection for each of them.

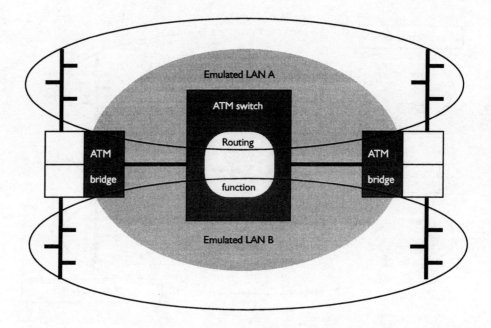

Figure 5.11 Interconnection of emulated LANs

If IP is used as the routing protocol, each virtual LAN has a unique IP network subaddress. A source station belonging to a virtual LAN first checks whether the destination station belongs to the same IP subnet as the virtual LAN. If it is not the case, the source station sends its packet to the default router that also belongs to the same virtual LAN. This router (or routing function, if integrated in the ATM switch) belongs to several virtual LANs and can thus forward the packet to the virtual LAN that connects the destination station.

In the strict LAN emulation context, it is not possible to set up a direct ATM connection between a source and destination stations which belong to different emulated LANs, and the forwarding through a routing function may limit the interworking performance.

IP and ATM

The LAN emulation provides the network layer protocols with a connectionless service based on a connection-oriented service. As it is transparent to the network layer protocols, it allows several of these protocols to be carried, and this constitutes a major advantage in a multiprotocol environment. However, if only one network layer protocol is used, a direct operation is possible between this protocol (IP, for example) and the ATM (sub)network. The address resolution then involves the association of IP and ATM addresses.

The protocol that allows such an automatic association of IP and ATM addresses works in a client server mode similar to the principles of the LANE protocol. This protocol was defined by the IETF (RFC 1577); in its simplest form, it covers a single IP subnet called LIS (logical IP subnet). A LIS is composed of clients and of a server in charge of the address resolution (**ATMARP server**). This server requests the IP and ATM addresses from each new client at connection time. Later, when a client wants to send a packet to a destination station located on the LIS, it asks the server for the associated ATM address, before setting up a direct virtual connection with that station.

This mode, also called **classical IP over ATM**, covers the communication needs for stations belonging to the same LIS. The location of interconnecting stations in two different logical subnets leads to the establishment of a virtual connection in each subnet and to routing the packets by means of a router which belongs to both subnets (Figure 5.12). The reassembly function of cells into packets, needed within the router, reduces the transfer rate significantly. The IETF proposes a more efficient protocol called next hop routing protocol (NHRP) that allows the establishment of a direct virtual connection between stations belonging to two logical IP subnets (see Figure 5.13).

A weak point in the classical IP over ATM mode came from the lack of a protocol between the servers, which thus knew only the clients belonging to their own IP subnet. By equipping the address resolution servers with an NHRP-type interdomain exchange protocol, it becomes possible to set up a direct virtual connection between the source and the destination stations, regardless of the number of subnets involved. If the server (called next hop server or NHS) cannot associate an ATM address with the IP destination address given by its client, it knows which server can make this association, by means of the interdomain routing protocol that allows the interconnected servers to keep their tables updated.

To support packet broadcast, new servers are being standardized: MARS (multicast address resolution server) and MCS (multicast connection server).

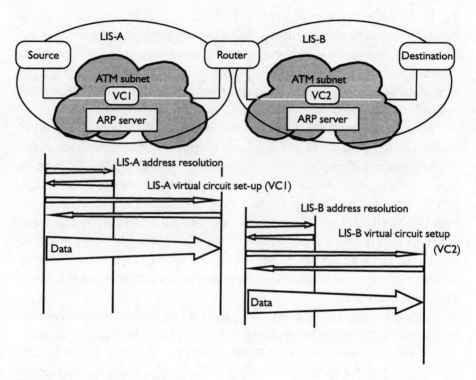

Figure 5.12 Classical IP over ATM mode

Multiprocol over ATM

The two protocols which allow the use of the ATM technology in a local area network environment suffer from some limitations:

- the LAN emulation presents all the characteristics of a LAN, including in terms of broadcast capability, but it does not currently allow the use of a direct ATM connection between stations belonging to different emulated LANs;
- the NHS allow a direct ATM connection to be established between two stations in different subnets, but this function is restricted to the IP.

To compensate for those drawbacks, the ATM Forum is working on a new protocol: multiprotocol over ATM (MPOA). Like its predecessors, it operates in a client server mode. The MPOA servers, called route server func-

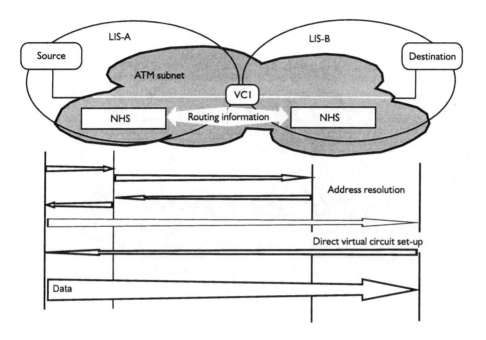

Figure 5.13 NHRP mode

tional group (RSFG), integrate two main functions brought by the multi-protocol routers:

- the automatic discovery of the network topology;
- the identification of the best route.

As far as the packet transfer function is concerned, it is provided by the MPOA clients, which are of two types:

- stations directly connected to the ATM network, called ATM-attached host functional group (AHFG);
- adaptation equipment between the ATM network and the conventional LANs, called edge device functional group (EDFG).

Other functional groups have also been defined, such as DFFG (Default Forwarder Functional Group) which, among other functions, is in charge of multicast operations.

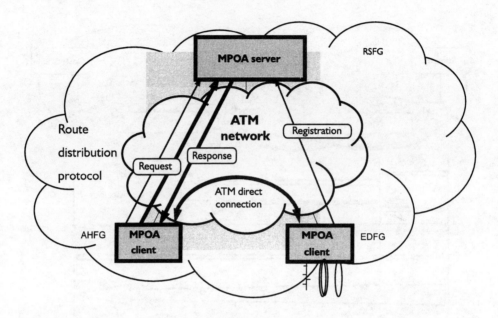

Figure 5.14 MPOA mode

When an MPOA client (Figure 5.14) receives or sends a packet that has to be switched at the layer 3 level on an ATM network, it is necessary to associate the destination network layer address (e.g. IP, IPX) with its ATM address, to allow the MPOA client to set up a direct connection with this destination. This address association is done by the MPOA server, which transmits the route parameters to the client using the RDP (route distribution protocol). This protocol is being standardized in the context of the ATM Forum. Some existing protocol elements, such as LANE, NHRP and MARS/MCS will be reused to speed up MPOA availability and reduce its complexity.

For some new applications, another approach could consist in developing an ATM-specific API (application program interface) that would keep some ATM advantages, particularly the qualities of service. The ATM Forum has started the definition of such an interface which authorizes a direct access to the ATM subnet and thus eliminates an intermediate layer such as LAN emulation or a minimum network layer function such as IP.

Appendix A

ATM standardization

ITU-T

The **ITU-T** (International Telecommunications Union-Telecommunication Standardization Sector) is one of the standardization bodies of the ITU (International Telecommunications Union). Its members are representatives of the member countries of the ITU (France Telecom for France) and some scientific and industrial organizations. The ITU-T was formerly called CCITT (Comité Consultatif International pour le Télégraphe et le Téléphone).

The ITU-T's standardization work is defined in plenary sessions, which are held every 4 years. It is then carried out by study groups (SGs) which are themselves made up of working parties (WPs). The group concerned with ATM is SG 13 (digital networks including ISDN) and, more especially, WP 8, which is responsible for broadband ISDN. ITU-T documents are called recommendations. The first ones concerning ATM were approved in 1988: they are Recommendations I.113 (Vocabulary of Terms) and I.121 (Broadband Aspects of ISDN).

ITU-T has defined ATM as the target transfer mode for broadband ISDN. A large number of recommendations relating to ATM have been approved or are about to be; the main ones are listed below:

I.113	Vocabulary of Terms;
I.121	Broadband Aspects of ISDN;
I.150	ATM Functional Characteristics;
I.211	Service Aspects;
I.311	General Network Aspects;
I.321	Protocol Reference Model;
I.327	Network Functional Architecture;
I.35B	ATM Layer Performance Aspects;
I.361	ATM Layer Specification;
I.362	AAL Functional Description;
I.363	AAL Specification;
I.364	Support of Broadband Connectionless Data;
I.371	Traffic Control and Resource Management;
I.413	User Network Interface (UNI);
I.432	UNI Physical Layer Specification;
I.555	Frame Relay Interworking;
I.580	General Arrangements for Interworking between B-ISDN and 64 kbit/s-based ISDN;
I.610	OAM Principles of B-ISDN Access;
Q.142x	Metasignalling;
Q.2010	General Introduction to Signalling in B-ISDN;
Q.2100	B-ISDN Signalling ATM Adaptation Layer, Overview description (see also Q.2110, Q.2120, Q.2130 and Q.2140);
Q.2500	B-ISDN Network Nodes, Introduction and Field of Application (see also Q.2510, Q.2520, Q.2530 and Q.2550);
Q.2610	Use of Cause and Location in B-ISDN and B-ISUP;
Q.2650	Interworking between B-ISUP and B-ISDN Signalling;
Q.2660	Interworking between B-ISUP and N-ISUP;
Q.2761	Functional description of the B-ISDN User Part of Signalling System No. 7 (see also Q.2762, Q.2763 and Q.2764);
Q.2931	B-ISDN UNI Layer 3 Protocol;
F.811	Service Description, Connection Oriented Data;
F.812	Service Description, Connectionless Data.

ANSI

Like AFNOR in France, ANSI (American National Standards Institute) is the United States national standardization body. It has a vast field of activity and delegates its authority to an appropriate body for each domain. ECSA (Exchange Carriers Standards Association) is the accredited body for telecommunications.

ANSI committee T1 is in charge of telecommunications. In particular, this committee issued the standards related to SONET (Synchronous Optical Network) acting on a proposition from Bellcore (Bell Communications Research). The T1S1 technical group is more especially concerned with high-speed networks.

The standards concerning broadband ISDN are usually the equivalent of the ITU-T recommendations (see the list below):

T1.105	Synchronous Optical Network;
T1E1.2/92-020	UNI PMD Specifications;
T1S1/92-185	UNI Rates and Formats;
T1S1.5/92-002	ATM Layer Specifications;
T1S1.5/92-003	AAL 3/4 Common Part;
T1S1.5/92-004	AAL for CBR (Class A) Services;
T1S1.5/92-005	Support of Connectionless Services;
T1S1.5/92-006	Services Baseline Document;
T1S1.5/92-007	Generic Flow Control (GFC);
T1S1.5/92-008	VBR AAL Service Specific Part;
T1S1.5/92-009	Traffic and Resource Management;
T1S1.5/92-010	AAL 5 Common Part;
T1S1.5/92-029	OAM Aspects (Technical Report);
T1S1.5/92-xxx	AAL Architecture for Class C/D and Signalling.

ANSI committee X3, which is in charge of computer networks and interfaces, is also heavily involved in ATM. Its X3T9.5 subgroup set the FDDI standard; similarly, the documents relating to HIPPI (high-performance parallel interface) and to FCS (fibre channel standard) were published by the X3T9.2 subgroup.

IEEE

This organization (Institute of Electrical and Electronics Engineers) is well known for its local area network standards, formulated by Project 802. IEEE group 802.6 issued the DQDB protocol standard for metropolitan area networks: it uses cells of the same size as ATM cells. For high-speed transmission, 100BaseT and 100VG-AnyLAN technologies were standardized by committees 802.3 and 802.12 respectively. Further, the 802.9 group, responsible for the connection of terminals to local and long-distance networks, is also interested in the work on ATM.

ATM Forum

The ATM Forum was formed in 1991 and currently includes about 700 companies. Its main mission is to speed up the development and deployment of ATM products through interoperability specifications. The ATM Forum produces implementation agreements based on international standards; another goal is to fill the gap in specifications when standards are not available.

The ATM Forum specifications are prepared in several working groups:

- physical layer (specifications of electrical or optical interfaces at various rates);
- signalling (UNI 3.1, UNI 4.0);
- P-NNI protocol (IISP, P-NNI phase 1);
- B-ICI protocol (broadband intercarrier interface);
- network management (specifications of network management interfaces);
- traffic management (VBR, ABR, UBR modes);
- service aspects and applications (e.g. circuit emulation, multimedia services);
- LAN emulation;
- residential environment;
- multiprotocol aspects;
- security;
- testing (test specifications).

The following list gives the approved specifications as of February 1996:

af-intro-0001.000   ATM Forum Document Roadmap;
af-intro-0002.000   Introduction to ATM and ATM Forum;
af-intro-0003.000   Glossary of ATM Terms;

af-phy-0015.000	ATM PMD for 155 Mbps over Twisted Pair Cable;
af-phy-0016.000	DS1 Physical Layer;
af-phy-0017.000	Utopia;
af-phy-0018.000	Mid-range Physical Layer for UTP-3;
af-phy-0029.000	6.312 Mbps UNI;
af-phy-0039.000	Utopia Level 2;
af-phy-0040.000	Physical Layer for 25.6 Mbps over Twisted Pair;
af-phy-0043.000	Cell-based Transmission Convergence Sublayer for Clear Channels;
af-phy-0046.000	622.08 Mbps Physical Layer;
af-phy-0047.000	155.52 Mbps Physical Layer for UTP-3l;
af-uni-0010.000	ATM UNI 2.0;
af-uni-0010.001	ATM UNI 3.0;
af-uni-0010.002	ATM UNI 3.1;
af-uni-0011.000	ILMI MIB for UNI 3.0;
af-uni-0011.001	ILMI MIB for UNI 3.1;
af-uni-0012.000	Differences between UNI 3.0 and UNI 3.1;
af-pnni-0026.000	Interim Inter-Switch Signalling Protocol;
af-bici-0013.000	B-ICI 1.0;
af-bici-0013.001	B-ICI 1.1;
af-bici-0013.002	B-ICI 2.0;
af-nm-0019.000	Customer Network Management for ATM Public Network Services;
af-nm-0020.000	M4 Interface Requirement and Logical MIB;
af-nm-0027.000	CMIP for M4 Interface;
af-saa-0031.000	Frame UNI;
af-saa-0032.000	Circuit Emulation;
af-saa-0049.000	Audio/Visual Multimedia Services: Video on Demand;
af-lane-0021.000	LAN Emulation over ATM 1.0;
af-lane-0044.000	LAN Emulation Client Management;
af-test-0022.000	Introduction to ATM Forum Test Specifications;
af-test-0023.000	PICS Proforma for DS3;
af-test-0024.000	PICS Proforma for STS-3c;
af-test-0025.000	PICS Proforma for 100 Mbps MMF (TAXI);
af-test-0028.000	PICS Proforma for ATM Layer (UNI 3.0);

af-test-0030.000	Conformance Abstract Suite for Intermediate Systems (UNI 3.0);
af-test-0035.000	Interoperability Test Suite for ATM Layer (UNI 3.0);
af-test-0036.000	Interoperability Test Suite for DS3, STS-3c, 100 Mbps MMF (TAXI);
af-test-0037.000	PICS Proforma for DS1;
af-test-0041.000	Conformance Abstract Suite for End Systems (UNI 3.0);
af-test-0042.000	PICS for AAL5;
af-test-0044.000	PICS Proforma for 51.84 Mbps;
af-test-0045.000	Conformance Abstract Suite for Intermediate Systems (UNI 3.1).

Several other documents are currently in progress, the main ones being the following:

LANE V2.0 Server-to-Server Interface;
Traffic Management 4.0 (ABR);
P-NNI V1.0 (Signaling and Routing);
UNI Signaling 4.0;
MPOA V1.0.

ETSI

The aim of ETSI (European Telecommunication Standards Institute) is to draw up European standards called ETS (European Telecommunication Standards). Whenever possible, these documents use existing standards as their basis (for example, ITU-T recommendations) and, once they are approved, they apply to all the member countries of ETSI. The following list gives the correspondences between the ITU-T recommendations and the standards approved by the members of ETSI:

ETS 300.298	⇒ I.361	ATM Layer Specification;
ETS 300.299	⇒ I.362	AAL Functional Description;
ETS 300.300	⇒ I.363	AAL Specification;
ETS 300.301	⇒ I.432	UNI Physical Layer Specification.

Other documents related to ATM include:

DE/NA-52617	AAL-1;
DE/NA-52618	AAL-3/4;
DE/NA-52619	AAL-5;
DE/NA-52206	OAM;
DE/NA-52209	OAM;
DE/SPS-5024	UNI Signalling (basic call);
DE/SPS-5034	UNI Signalling (supplementary services);
DE/SPS-5026-1 and 2	SAAL;
DE/NA-53204	Frame relay service (FR/ATM);
DE/NA-53205 and 6	High-speed connectionless service (CBDS/ATM).

The table on page 120 gives the addresses of the main bodies working on ATM standardization.

The text of the ATM Forum documents is available through the World Wide Web server called **www.atmforum.com**. Similarly, it is possible to obtain the text of the ITU-T recommendations using a connection to the **info.itu.ch** server.

Organization	Mail address	Telephone	Fax
ATM Forum	2570W.	(+1) 415 949 6700	(+1) 415 949 6705
	El Camino Real, Suite 304		
	Mountain View		
	CA 94040-1313, USA		
ETSI	Route des Lucioles	(+33) 92 94 42 00	(+33) 92 65 47 16
	F-06921 Sophia Antipolis		
IEEE	445 Hoes Lane	(+1) 908 981 0060	(+1) 908 981 9667
	PO Box 1331, Piscataway		
	NJ 08855-1331, USA		
ITU-T (ex CCITT)	Place des Nations	(+41) 22 730 6666	(+41) 22 730 5337
	CH-1211 Geneva 20		

Appendix B

Families of ATM products

This appendix divides the ATM products into five basic categories. This proposal commits only the authors and must not, under any circumstances, be seen as being linked to a product strategy.

As indicated on Figure B.1, the market of the ATM products can be split into several segments:

- the products to be used within the telecommunications network operators' backbone (also called **core network**). These are high-capacity switches (more than 50 Gbit/s) which only offer ATM interfaces up to 622 Mbit/s. They are usually provided by the manufacturers of public network switches;
- the equipment constituting the access network between the operators' core network and the local area networks. These multiservice switching platforms, which may be located near the user's equipment, use the ATM technology as switching means (up to 10 Gbit/s) and propose a wide set of service interfaces (e.g. frame relay, CBDS/SMDS, emulated G.703 circuits, ATM virtual circuits). For this type of products, the manufacturers are often different from the ones which provide the backbone switches;

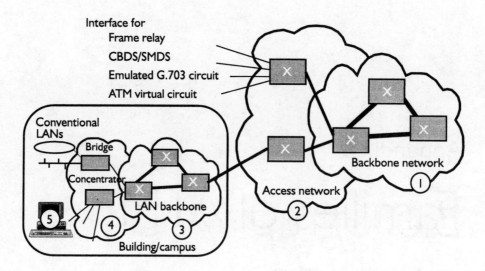

Figure B.1   Market segments for ATM products

- the products which form the backbone of the campus network. These are ATM switches characterized by a medium-size capacity (10 Gbit/s, for example) and limited to the local area environment. They are provided by a third category of manufacturers, coming from the market of LAN routing and switching products;

- the equipment allowing access to that campus network backbone, i.e. bridges connecting the ATM local area network to the conventional LANs (**LAN switches**) or ATM concentrators (**workgroup switches**). They are proposed by a fourth category of manufacturers, which usually come from the market of bridging and switching devices for the LAN environment, and which partially overlaps the previous category;

- the products that allow ATM technology to be brought to the workstation. They usually consist in interface cards and associated software modules. That fifth category of providers also overlaps the previous ones.

Bibliography

Books

Armbruster, H. *The Flexibility of ATM*, Proceedings of the ATM conference (Paris, April 1993).

Boisseau, M., Demange, M., Munier, J.-M. (1995) *Réseaux haut débit*. Eyrolles.

Cuthbert, L.E., Sapanel, J.C. (1993) *ATM: The Broadband Telecommunication Solution*. Institute of Electrical Engineers.

Coudreuse, J.-P. *et al.* (1991) *Spécial ATM*, L'Echo des recherches no. 144/145.

de Prycker, M. (1990) *Asynchronous Transfer Mode: Solution for Broadband ISDN*. Ellis Horwood, Chichester.

Handel, R., Huber, M.N. (1991) *Integrated Broadband Networks: An Introduction to ATM-based Networks*. Addison Wesley, Wokingham.

Kyas, O. (1995) *ATM Networks*. International Thomson Publishing, London.

Pujolle, G. (1995) *Les réseaux*. Eyrolles.

McDysan, D., Spohn, D. (1994) *ATM Theory and Applications*. McGraw Hill, New York.

Onvural, R. (1994) *Asynchronous Transfer Mode Networks: Performance Issues*, Artech House.

http://www.canarie.ca/ntn
Bill St Arnaud has gathered information concerning the ATM networks on the CANARIE (Canadian Network for the Advanced Research, Industry and Education) project's home page. Note that the FAQ (frequently asked questions) rubric is available in both English and French.

http://www.nic.surfnet.nl/surfnet/projects/atm/index.html
On the Surfnet (the Dutch research network) project's home page, Victor Reijs and Gett-Henk Bakker have gathered general information on ATM deployment in Europe and in the United States.

In addition, access to the Internet allows the text of the ITU-T recommendations and ATM Forum's specifications to be obtained (p. 120).

Index of abbreviations

Thematic index